# Restoring, Tuning & Using Classic Woodworking Tools

## Michael Dunbar

*Photos by Andy Edgar*
*Drawings by Barry & Tracy Kane*

Sterling Publishing Co., Inc.   New York

# Acknowledgments

I wish to thank the following people for their help in writing this book: Don Baldwin, Ernie Conover, David Draves, Tom Hinckley, Charlie LeBlanc, Malcolm MacGregor, Larry McManus, David Sloan, and Carl Dohn Jr.

Edited by Michael Cea

**Library of Congress Cataloging-in-Publication Data**

Dunbar, Michael.
    Restoring, tuning & using classic woodworking tools / Michael Dunbar.
        p.    cm.
     Bibliography:  p.
     Includes index.
     ISBN 0-8069-6670-X
     1. Woodworking tools—Maintenance and repair.  2. Woodworking tools—History.  I. Title.  II. Title: Restoring, tuning, and using classic woodworking tools.
TT186.D863     1989
684′.08′028—dc20                    89-35456
                                                CIP

# Contents

# Introduction

Every woodworker that I have ever known—whether he or she works by hand or totally with machines, is an occasional woodworker or a professional—eventually acquires some tools that once belonged to someone else. They might be planes that belonged to his father or chisels that belonged to his grandfather. One way or the other, every woodworker will find himself owning at least a few such tools; most woodworkers own a lot of them.

I am very partial to classic tools, tools that in this book are referred to as second-hand. Second-hand tools, which were built in the finest tradition of craftsmanship as far back as the 18th century and can be found generally in large numbers today on the marketplace, play essential roles in today's fine-quality woodworking shops. I have been using them since I began woodworking in 1971. My affection is not just sentimentality, nor is it a desire to be quaint. Using older tools is a very practical way to work wood. They are good tools, and when compared to the hand tools that are being made today, are most often the better ones.

In the following pages I explain how to buy, recondition, tune, and use most of the second-hand tools you will commonly find. Most of this information will also be helpful to you when you buy a new tool, as few tools come ready to use right from the box. They always have to be tuned and sharpened, procedures that incorporate many of the steps that I describe for reconditioning older tools.

This book is a practical guide. In writing it, I assumed that you already have a basic knowledge of woodworking, as well as a familiarity with hand tools—their purposes, how they are used, etc. Since it is a practical guide, it is intended for the woodworker rather than the collector. As a result, it does not have price lists or check lists of toolmakers.

Although the purpose of this book is to explain how to find, recondition, and use second-hand tools, not every type of tool is discussed, for two reasons. First, although I have done all sorts of woodworking, ranging from inlay to timber-framing, I have not worked with every tool found on the second-hand market and cannot speak with any authority about them all. For example, I only use a wooden spokeshave. I have tried metal ones, and so know how they work (I also know that they do not perform as well as their wooden counterparts). So, although they are a common second-hand tool, they are not included in this book.

Secondly, I live in New England, which was settled much earlier than other areas of the United States. Consequently, some of the tools found in New England are much older than those commonly found in other areas. Because many of these very old tools are not commonly found everywhere, I have excluded many (but not all of them).

If by chance a tool has been left out that is of interest to you, I am confident that the information that is supplied will allow you to infer what you need to know to recondition, sharpen, and use it.

Although I do not generally recommend using what are usually called antique tools, and explain why on page 17, I have included some. Wooden planes are a good example. So many dealers now sell these tools that they can be purchased by mail order even if they are not regularly found where you live. So, no matter where you are located you can easily obtain wooden planes. Many of them are still very serviceable.

Each chapter deals with a different category of tools, and begins with a general discussion of that type of tool that includes the following: the tool's history and evolution, its purpose, how it works, and the features and characteristics that make it worth purchasing. Also discussed are ways to recondition and tune that particular tool, and the problems that (in my experience) are most common to that type of second-hand tool. A section on selecting the tool contains information about the kind of damage (specific to that tool) that cannot be repaired or would be either so complicated or time-consuming that buying the tool would prove impractical.

Although Chapter 8 is a general description of tool-sharpening, any special sharpening problems that can be encountered with a particular tool are discussed in its specific chapter. And, finally, no discussion of the tool would be complete without information on how to use the tool.

There are two basic sources of written information available to those interested in learning about second-hand tools. Both proved to be of great help to me when I was learning about hand tools and hand woodworking. The first is tool catalogues. Since as early as 1816 (when Smith's *Key to the Various Manufactories of Sheffield* was published), tool makers have been publishing illustrated catalogues of their tools. Several pages from both these catalogues and general catalogues have been printed in this book.

The other source of woodworking written material is, of course, woodworking books. For centuries, enterprising woodworkers have been publishing books explaining the how-tos of woodworking. However, there is one essential difference between the books written today and those written in the latter part of the nineteenth century and the early part of this one: The earlier books show how to work by hand using hand tools.

In the bibliography on page 254, I have listed the names of some tool catalogues and woodworking books available in reprint. If you find that you enjoy working with second-hand tools, perhaps you should read some of these.

A few terms appear throughout this book that should be clarified here. The words *iron* and *cutter* have the same meaning and are used interchangeably. *Tuning* refers to all the processes needed to prepare a tool for use. Sharpening and adjusting are part of this process. For example, flattening the sole of a plane and bevelling the underside of a drawknife would be considered part of the tuning process. Both old and new tools have to be tuned before they will give you their best performances.

*Reconditioning* (or restoring) is only done to second-hand tools. This process includes cleaning, as well as making repairs and replacements. After a second-hand tool is reconditioned, it still has to be tuned.

# Tool Buying Guidelines

# 1
# Why Buy "Classic" Tools?

The tools described in this book are some of the best hand tools ever made, often superior to what is sold today. This is because only a generation ago woodworkers were much more reliant on hand tools than they are today. The relative cost of machinery was much higher, so even the owner of a small professional woodworking shop was completely dependent on his hand tools. They had to be of the best quality he could afford. If the tool were not capable of doing the job, he was out of work.

At that time, tool manufacturers were proud to produce hand tools of the quality needed by woodworking professionals, just as today the makers of better-grade powered tools are proud of their products. Over the decades, as less-expensive power tools were developed, the demand for quality hand tools was gradually reduced. In some cases, manufacturers were forced to eliminate their best lines altogether.

There are some notable differences in quality between tools made in the latter half of the 20th century and those made earlier. Hand saws clearly illustrate this difference. Compare the saw handles shown in Illus. 1. The handle shown at the bottom of the photo was made either during the late 19th century or early in the 20th century. It is a very handsome tool, almost sculpture-like. But besides its obvious aesthetic qualities, it is also quite functional. All of its edges have been gracefully rounded, which means you could use it for an extended period of time without raising a blister.

The saw is also balanced so that the placement of its handle permits the most efficient use of muscle power and allows the user to lock his wrist

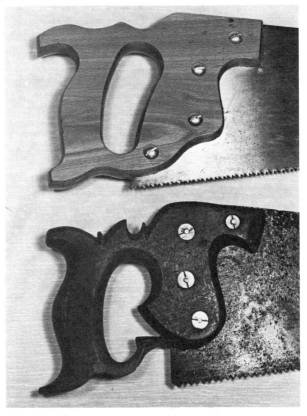

*Illus. 1. These two saw handles show how tool design has changed for the worse. The lower saw handle is a type that is commonly found on the second-hand market. It is sculpted, so it is not only more comfortable to hold, it is also more pleasant to look at. Note, too, that it is mounted at an angle that makes it most efficient to use. The other saw is only several years old. Its handle has sharp edges that are uncomfortable to grip. At least it is made of solid wood. Many saw handles are now made of plywood and even plastic.*

in a position that is comfortable and conserves strength. Today, the application of these principles is called ergonomics. This is a recent science, but, ironically, there is evidence that older tools show more concern for ergonomics than do those made today.

The second saw handle shown in Illus. 1 was purchased at a hardware dealer and is typical of what you will find today. Very little effort was made to relieve the edges, and you would develop blisters if you used it for a long period of time. The manufacturer did not have to make it comfortable to work with because today's buyer would normally use it infrequently. Most new hand saws are not sold to woodworkers, but to home owners who need to cut an occasional board or do small projects around the house.

A professional or an active nonprofessional does most cutting with powered saws, and, with the exception of some joinery (for example, dovetails), many woodworkers have eliminated hand saws from their work. Since there is little demand for high-quality saws, it is not surprising that they are not made in large quantities.

Illus. 2 also contrasts the differences in quality between new and second-hand tools. Shown are two metal bench planes. One is an old Bailey Patent #4 smooth plane; the other was purchased from a hardware dealer. The castings on the old plane, which bears a 1902 patent date, are much more refined and more carefully finished than those on the new one. The tote (handle) on the new plane is shapeless and shows only the slightest indication that it was meant to be held in a human hand. It was obviously not intended to be used all day long.

Second-hand hand tools are not only better made than new ones, there is also more of a variety of them. Formerly, manufacturers sold many more types of hand saws than they make today. A hand woodworker needed different saws for cutting curved work, ripping, coarse crosscutting, fine crosscutting, and for carcass construction and dovetailing. Visit a hardware dealer, or even a tool store. It is very difficult to find a specialized saw such as a 12-point crosscut saw or a 7-point ripsaw.

Besides being often nicer to look at, more comfortable to use, and of a greater variety, older tools have another very important advantage over new tools. Many of the second-hand tools were made at a time when there was a strong tradition of craftsmanship.

Those who made the tools also worked with them. As craftsmen, they understood what makes a tool so effective—and applied this knowledge when making the tool.

New tools, on the other hand, are designed by engineers who want to make the tool as inexpensively as possible. Very few of these people are themselves craftsmen, and some do not even understand how the tool works. Also, many of

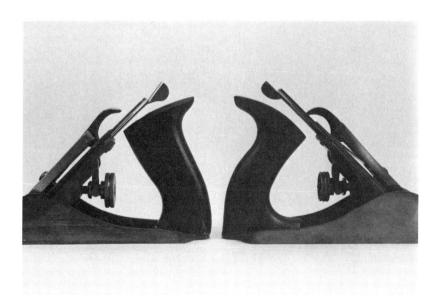

*Illus. 2. Compare the handle on the Bailey #4 smooth plane at right (made between 1905 and 1914) with that on the much newer plane. Note also the lateral adjustment lever. On the new plane it is merely stamped from sheet metal. There are also differences in the quality of the castings, but these cannot be easily seen in a photograph.*

today's hand tools are made by machines run by machine operators, most of whom know very little about working wood. As a result of these factors, some new tools are being sold that have major flaws that prevent them from working well, and, in some cases, from working at all. This has been borne out continuously in the woodworking classes that I teach throughout the United States. Students in these classes often buy tools with such problems.

# 2
# Basic Tool Information

## Historical Overview

Though a brief history of each type of tool is given in that particular chapter, it would be helpful for you to have a general perspective of the history of hand tools.

Woodworkers have always made some tools for themselves, and they still do. However, during the 18th century the production of such tools as planes became a specialty. Master plane makers worked by hand, aided by journeymen and apprentices. These men were craftsmen and worked in what is called the craft-shop tradition.

During the Industrial Revolution, tool factories were built, and, as toolmaking became industrialized, the craft-shop tradition died out. Thus, the history of toolmaking can be divided into two distinct periods: the preindustrial or craft-shop period and the industrial period. Woodworking machinery was also developed during the Industrial Revolution, so woodworking can be divided into these two periods as well. In the preindustrial period, craftsmen worked by hand with their journeymen and apprentices. During the industrial period, much woodworking was done in factories with machines, and work that was still done by hand was done with factory-produced hand tools.

The Industrial Revolution did not occur overnight; there was a period of transition. At first, tool factories continued to make tools patterned after those made in the craft-shop tradition—for example, wooden planes. After the American Civil War, many tools that had been made of wood were being made of cast iron and steel. These factory-made tools are the ones you will usually find on the second-hand tool market. Earlier craft-shop (preindustrial) tools are of interest to collectors and, as a result, their prices are often high.

In America, you will find many English tools from the preindustrial and early industrial periods. Lately, many American tool dealers have made annual buying trips to England, and have imported second-hand tools by the container load. As a result, many of these tools are recent arrivals.

On the other hand, from the earliest times foreign tools (mostly English) were available in America and were in widespread use. Woodworkers who could afford them favored English tools because they were superior to domestic products. For this reason, some of the second-hand English tools sold in America have been here for a long time.

After the American Civil War, American toolmakers started to establish international standards. Inventors like Leonard Bailey, Edward Hyder, and William Henry Barber developed tools that were accepted equally well in England and in America. English woodworkers once referred to these tools as "American pattern" tools.

Recently, some continental and Scandinavian tools have appeared on the American market, indicating that dealers are now buying in these countries. Coincidentally, some English dealers now make buying trips to America to purchase and repatriate English tools.

## General Types of Tools

Following is some basic information you should know about the general categories of second-hand tools before you consider buying them.

### "Best" versus "Common" Tools

Toolmakers used to make not only a greater variety of hand tools than they do now, they also made different grades of tools; this is similar to a manufacturer today offering four or five different grades of cordless drills. Toolmakers usually had a high-priced line that they would advertise as their "best" tools. These tools were not just those of the highest quality, they also had the greatest number of features, just as the most expensive cordless drill has the most options. In the past, woodworkers were very concerned with the way their tools looked and, as a result, best-grade tools are usually very handsome. They were often made of more expensive materials, such as brass and exotic woods..

"Common" was the term used for the lesser-grade tools. Common tools were usually unadorned and were made with less-expensive materials such as iron and native hardwoods. However, common tools usually work just as well as the "best" tools. For the modern woodworker who wants to use secondhand tools, the price, not performance, usually distinguishes best from the common grade. Tool collectors generally prefer best-grade tools, and they will pay higher prices for them. Still, best-grade tools do occasionally surface at bargain prices.

# Gentleman's versus Handyman's Tools

Woodworking has always been a popular hobby. During past centuries, toolmakers made tools for serious (but non-professional) woodworkers. These tools were called Gentleman's tools (sometimes shortened to Gent's tools). They were of good quality, but tended to be smaller and lighter in weight than those used by tradesmen.

Around the turn of this century, toolmakers began to make tools that could be used in odd jobs. They were called Handyman's tools. Unlike the older Gent's tools, Handyman's tools were poorly made. Many of them can be found on the second-hand market, and I advise against buying them. They were never intended for serious woodworking, and you will be disappointed with them.

Handyman's tools are still made. You can buy them new in hardware shops and even in department shops. Many of them are made in developing countries, and they are still of low quality. Some of these new tools have also found their way into the second-hand market, and I advise against buying them as well.

# 3
# Buying Guidelines

Only two decades ago, all wooden planes were very inexpensive. Boxes of them (regardless of function or condition) could be purchased very cheaply. In fact, stories were frequently told about auctioneers who sold boxes of moulding planes as stove wood because they were of so little value.

This is no longer the case. Now, there are people who collect wooden planes as a hobby, and even for investment purposes. As a result, the pricing of these tools has become very sophisticated. The value of an individual plane is now determined in the same manner as investment-grade coins, by the following criteria: maker, condition, age, and type. Though ordinary planes are still inexpensive, the most sought-after ones can cost a considerable amount of money.

Take note of how dramatically the prices of wooden planes have changed in two decades. It is possible that other second-hand tools will follow a similar pattern. Even now, people are specializing in collecting just Stanley planes.

If you are thinking of outfitting your woodworking shop with second-hand tools, first decide which has a higher priority, time or money. Second-hand tools are not retailed in the same manner as new tools, and you have to spend more time looking for them. However, because these tools are second hand, they often cost less than a comparable modern tool. This factor, coupled with the fact that these tools are usually better made, makes them a real bargain.

Buying second-hand tools can be a new experience, as the market is still unorganized. As it develops, the difference between it and the retail market for new tools will become less pronounced, but, for a number of reasons, the two will never be the same.

When I began my career as a woodworker, old tools were found at yard sales, flea markets, and antique shops. They can still be found in these places. However, during the last couple of decades, people have begun to deal exclusively in old tools. Now, in most parts of the United States and England there are shops that are either entirely stocked with second-hand tools or where such tools make up a large percentage of the merchandise. (See Illus. 3.) In this book, I refer to these businesses as specialized tool dealers. Some tools I will describe are found almost exclusively in such specialized shops.

These specialized tool dealers are becoming more and more sophisticated in their operations. For example, some of them now sell by mail order. They issue catalogues on either a regular or semi-regular basis and have continuously expanding mailing lists. Unlike new tool catalogues, second-hand tool catalogues are not usually of great quality. Catalogues for new tools do not have to change very much, issue to issue, and so expensive printing plates can be reused. On the other hand, a second-hand tool dealer's merchandise might change completely between

*Illus. 3. Second-hand tools are available throughout the United States, not just on the east coast. This extensive inventory is just one corner of Andy Anderson's shop on Entrada Drive, in Santa Monica, California. (Photo courtesy of Don Baldwin.)*

*Illus. 4. Bristol Design on Perry Road, Bristol, England is one of the largest antique and second-hand tool dealers in the United Kingdom. The company publishes a regular catalogue and will ship. It has a large number of subscribers in the United States.*

catalogues, requiring each catalogue to be completely different.

Supply and availability are important differences between old tools and new ones. A new tool dealer simply orders a supply of a particular tool that approximates the number that can be sold in a reasonable amount of time. If the store runs out of stock, it places your request on back order and tells the supplier to send another shipment. That is not possible with second-hand tools. Because the supply is limited to the items a dealer can find, a particular tool is not always in stock. You will learn to make mental lists of what tools you need and to buy those tools when you find them.

The law of supply and demand usually controls the prices of second-hand tools. Common tools, ones that you see often, are usually less expensive than tools that are hard to find.

Just as supply and demand affect the price of old tools, so does their condition. Second-hand tool dealers are usually very careful when describing a tool. They try to be very accurate, sometimes drawing more attention to a tool's flaws than to its virtues. Like rare coin dealers, some tool dealers have even tried to create objective indexes by which their tools can be graded. The system shown in Table 1 was developed by the *Fine Tool Journal*. It is called The Standard Condition Classification System and is used by many established tool dealers.

The following are typical descriptions of second-hand tools. They are quoted exactly as they appeared in a recent mail-order tool catalogue. I have omitted the prices.

"STANLEY #113 CIRCULAR PLANE, type "AA" mark on the iron and type "MM" lever cap, in excellent condition. G+."

"#55 UNIVERSAL COMBINATION PLANE, S. W. Hart, complete with all 55 cutters, in original box, missing only original lid. A new instruction manual will be furnished, needs heavy cleaning. G."

"Jack Plane w/lignum vitae sole—attached with 10 countersunk screws & wood plugs (one missing). Looks Scottish and iron is a Mathieson. Dirty, split cheek, chip tote, wedge beat up. Solid. Gd."

"Transition Jointer 24." As found, will clean to Good+."

"14″ Sash Brass backed (heavy) back saw 11 pt. JOHN COCKERILL, Sheffield. Hdl is closed and flat w/trivial dings. Bld has no rust or pits but teeth need recutting. *Vy* attract! Gd−"

"Dado J. KELLOGG, Amherst, Mass 1835–67. Brass thumb screw & plate for adj. stop (works fine). Repl wedge for spur cutter. Two tiny fence holes. Iron & Cutter mint. Fine."

Most catalogue tool dealers will allow you to return items that do not meet your satisfaction. However, before buying a tool, ask about the dealer's return policy. Since these dealers also sell to tool collectors, explain that you are a woodworker and want to use the tool—in other words, that you are more interested in how well it will work rather than its "collectibility."

There is another reason why it is important to tell a dealer you are a woodworker. Many second-hand tools were in production for many years, and during that time changes and improvements were made to them. For example, when Stanley first introduced its very useful #113 Circular Plane in 1879, it did not have a lateral adjustment lever. This feature, along with some other improvements, was added in 1898, making the #113 an

| Category | Wear | Finish | Usable | Repairs | Rust | Misc. |
|---|---|---|---|---|---|---|
| New | None | 100% | Totally | None | None | +Orig. Pkg. |
| Fine | Minimal | 90%–100% | Totally | None | Trace | |
| Good+ | Normal | 75%–90% | Yes | None/minor | Minor | Some dings & scratches |
| Good | Normal | 50%–75% | Yes | Minor | Minor | Small chips |
| Good− | Normal | 30%–50% | Probably | Minor | Minor–Mod. | Small chips & cracks |
| Fair | Heavy | 30% | No | Moderate–Major | Moderate–Major | |
| Poor | Heavy | N/A | No | Major | Major | |

*Table 1. Grading system developed by* Fine Tool Journal.

even better tool. Even though collectors are most interested in the earlier model, as a woodworker you want to use the later, improved version.

If you are looking for a particular tool, do not hesitate to shop by telephone. Maintain a list of telephone numbers of specialized tool dealers. You do not have to limit yourself to dealers in your area. Most dealers ship their tools, so they can be located anywhere in the country, or even overseas. Some dealers keep lists of tools their customers want to buy, and call when they have the item in stock.

Auctions are one way of buying tools that appear exclusively on the second-hand tool market. Auctioneers who specialize in selling antique and second-hand tools regularly hold auctions that are big enough to be advertised in major newspapers. However, auctions present a problem for the woodworker who wishes to use these tools. A good auction collection will usually contain a large percentage of expensive antique tools, and these auction sales regularly set price records. It is necessary for the auctioneer to include many high-priced tools since they help to pay the auctioneer's overhead.

However, an expensive collection does not necessarily rule out the possibility of finding second-hand tools at a bargain. Many tool auctions have become major events, and the surrounding atmosphere attracts many dealers. Some dealers set up booths and sell good, serviceable second-hand tools. Collectors and woodworkers even sell tools to each other out of the trunks of their cars.

If you want to attend auctions and buy mail order from specialized tool dealers, first you have to get your name on their mailing lists. Begin by contacting one or more dealers or auctioneers and asking them to include you in their next mailing. Because mailing lists are usually bought and sold, your name will be added to many more lists.

Remember, however, unless you make an occasional purchase, the computer will eventually eliminate your name.

The specialized dealer's shop, auctions and mail-order sales make up the established and organized part of the second-hand market. However, many second-hand tools are available in other places. Old tools are still regularly bought and sold by second-hand stores, pawn shops, and at yard sales and flea markets. Since the difference between second-hand tools and antique tools is somewhat blurred, old tools are also a staple in many antique shops.

Although I do buy some tools through the mail, I still enjoy a good tool hunt. I like to take a day and drive to several shops and flea markets. Since I never know what I am going to find, I always keep a list of the tools and sizes I need.

When shopping for tools, take along a kit. This kit should contain some basic tools such as a screwdriver and a pair of pliers, which are needed to disassemble tools for inspection, as well as a tape measure, a small try square, and a magnifying glass. It should also include some basic reference books with which you can identify what you have found and determine whether or not all the original parts are included with the tool. I am hopeful that this book will become part of your kit.

One final word about the second-hand tool market: When you purchase a new tool, there is generally no opportunity to make a deal with the company that owns the store. The tool's price is the price on the tag. The store's attitude is usually "take it or leave it." On the second-hand market, prices are often much more flexible. Never be afraid to suggest to the dealer that the sale be structured in an unusual manner—for example, a trade. Most dealers are use to making deals, and the worst he or she can say is no.

# 4
# Selecting the Proper Tool

It is important to understand which tools should be bought and which shouldn't. Because tools vary in quality, many are not worth owning. In fact, you must be careful and very selective. The majority of second-hand tools are, for one or more of the following reasons, unsuitable for use.

When buying second-hand tools, it is usually advisable to purchase only those that are in excellent condition. Whenever possible, avoid tools that need repair and those that are heavily worn and have little working life left in them.

Of course, the amount of work you are willing to do to a particular tool is determined by how badly you want or need it. This is a subjective judgment that only you can make. The tool may be worth restoring if it is very rare, is something you need or has a strong sentimental value. On the other hand, if the tool is common and you have no attachment to it, find one that does not need repair.

Following are some of the types of tools you should *not* buy to use for woodworking.

## Antique Tools

Beware of tools with prices that are the result of their appeal to collectors, or that have a high value as antiques. The word antique implies great age, but is only one of many factors that make a tool interesting to collectors. Some very old tools have less value than others that are much younger.

Antique tools have not only become collectible, they have also become a means of investment. As with fine art, antique furniture, and coins and stamps, novices should be very careful when buying antique tools. If you do not understand the circumstances that determine value, you could lose a great deal of money. It is very possible to pay much more for an item than it is worth—much more than you would receive if you decided to resell it.

Once you own an antique tool, be very careful as to how you use it. If the ordinary tuning that is necessary before any tool (old or brand new) can be put to use is not done properly, it can lower the value of an antique tool.

There is one final reason why woodworkers should leave antique tools to the collector: their recent high prices have inevitably attracted dealers selling fake "antiques." In this market a little knowledge is dangerous. The novice soon hears of the high prices paid for rare tools, but he at this point has not yet learned how to identify a real antique and may purchase a fake that is being sold at what seems a bargain price.

In the first chapter, I mentioned the advantages of second-hand tools; they are usually of better quality, and often less expensive. These characteristics do not necessarily apply to antique tools. So, unless you want to collect antique tools and invest the time necessary to learn about them and their prices, stay away from them.

# Damaged Tools

Avoid tools that need major repairs. Repair work often has to be done by another craftsman who is capable of doing work that you yourself cannot do, for example, a machinist or a blacksmith. This means that the tool's real price becomes the sum of what you paid for it plus the cost of the repair. And even if you can do the work yourself, your time has value and has to be factored into the total cost.

Seldom does a repair make economic sense, although there are always exceptions. I recently made some repairs to a ⅜-inch wooden dado plane that had been purchased by a woodworker friend. The front of the sole (a narrow wooden strip, no wider than the dado it cuts) had been snapped off and was missing. He purchased the damaged plane for just under ten dollars, deciding not to buy an identical undamaged dado plane that cost just under 30 dollars. I made the repair by cutting a shallow rabbet and gluing in a new piece of ⅜-inch beech. The repair took about 15 minutes and saved him a considerable amount of money.

# Tools with Missing Parts

Most second-hand tools have not been used for many years. In the meantime, they were either stored or left lying around. During this period of disuse, many tools lose removable parts. The more parts a tool has (such as a combination plane with all its separate cutters), the greater the chances that all the parts will not be with the tool when you acquire it.

If you do buy a tool that is incomplete, one solution is to duplicate the missing part. This is often more difficult and timeconsuming than making a repair. If the part is made of wood, you, as a woodworker, are more likely to be able to make it. Replacing a part that is made of forged or cast metal, however, requires skills that most woodworkers do not have.

Another solution is to find a replacement part. This means that you will not be able to use the tool until you find another identical, incomplete tool or a damaged tool that can be cannibalized. Once you have found this tool, you have to purchase it. The price of your one complete tool will be the sum of the two incomplete ones. The two purchases, plus the effort that went into finding them, seldom add up to a bargain.

Although I try to avoid buying incomplete tools, I occasionally find something that I want badly enough to justify ignoring my own advice. In anticipation of this possibility, I purchase loose items such as moulding plane irons, ferrules, saw nuts, and various types of handles, and stockpile them. Most tool dealers have a "dollar box" where anything in the box is sold at a fixed price, often one dollar. Loose parts are often tossed into the dollar box. I never leave a shop without taking a moment to paw through the box's contents.

Avoid any tool that has a badly chipped or cracked casting. However, sometimes a tool is so essential that such defects have to be accepted. For example, I own an old Bailey cast-iron bench rabbet plane (see Chapter 14). The casting and the machining on this plane are much finer than on the comparable plane that is made today. Unfortunately, the body was cracked just above the opening in the cheeks. The only craftsman I could find with the skill to repair it was a gunsmith. He was able to braze it so that it worked as well as when new. However, I did have to take the time to find someone who could do the repair and to deliver and pick up the tool; I also had to pay for the work. Had the plane not been such a good tool, I never would have invested that much time, effort, and money on it.

The decision to repair a tool or to replace a missing part is subjective and must be made by you. I will invest more time and effort into repairing a tool that I either need, is unusual, particularly appeals to me, or has sentimental value.

# Tuning, Cleaning, and Refinishing Tools

# 5
# Cleaning and Refinishing Tools

Most preindustrial wooden tools were probably not finished by the maker. However, over the years oil from the owners' hands has built up and created an unintended finish on the tool. During the mid-19th century, toolmakers began to apply either varnish or lacquer to wooden parts of tools. Cast-iron parts were coated with japan (a varnish made of linseed oil and asphalt pitch that is baked at 400 degrees F.). Brass was lacquered to prevent it from tarnishing.

Tool collectors are very concerned with what is called an "original finish," and a tool that retains all, or most of, this surface is usually worth much more than one that has been stripped or refinished. Thus (with some exceptions), I advise against stripping a tool, although the decision is up to you.

There are some circumstances where you have no choice but to refinish a tool. For example, sometimes a past owner has identified his tools by painting them. When I was growing up, I had a friend whose father covered every tool he owned with green paint, a sort of color-coded proof of ownership.

Paint also gets on old tools in accidental ways. Workshops are very often spattered with paint. Whether or not the paint was intentionally put on the tool, it looks sloppy, and, more importantly, will rub off on the wood you are working.

Old paint, grime, and the original finish can usually be dissolved with a premium grade of paint stripper. When I do strip the paint, I apply the stripper with an old paintbrush. After it has softened or lifted, I remove it with fine steel wool, usually grade 00 or 000.

I have refinished tools that I have had to repair, especially if the repair requires either the addition of a new wooden part or a patch. The old finish, suspended in the steel wool as a liquid grime, makes an excellent stain for disguising the new wood.

If you are considering stripping a tool, perhaps it would be wise to wait until you have gained more experience. You can always remove a finish. You can never put it back if later you decide you have made a mistake.

You do have to clean most second-hand tools before you can use them, but this is different from stripping them. Proper cleaning will not damage the finish. General procedure is to disassemble the tool and deal with the individual parts. Mineral spirits (paint thinner) is a good solvent and removes most dirt and grime. You can soak smaller parts or apply the mineral spirits with a brush.

# 6
# Heat-Treating Techniques

The ability to heat-treat steel is a handy skill for any woodworker. It is even more useful when you are reconditioning second-hand tools. I have had to heat-treat the following tools: a mortise chisel with a broken tip; edge tools that were inadequately hardened; tools that have been through a fire; and new irons for moulding and combination planes.

Do not try to teach yourself to heat-treat with a tool that you are reconditioning. First practice on something less valuable. You can heat-treat the blades or cutters of most tools with the following, easy-to-obtain items: a MAPP gas torch, a coffee can full of water, some 220 grit sandpaper, a new file, and a container of tightly packed wood ashes or lime. (See Illus. 5.)

*Illus. 5. You will need a MAPP gas torch, a can of lime, and a can of water (or oil for oil-hardening steels) to heat-treat a piece of steel.*

Heat-treating consists of three basic steps: annealing, hardening, and tempering. When annealing, you remove what remains of the old temper or any uncertain effects caused by accidental overheating, such as a fire or heavy grinding. You want to make the steel as soft as possible. Never skip this step because you assume that the steel is already soft. You may have disappointing results.

Before annealing, test the steel with a file, as this will allow you to contrast its hardness at various stages during heat treating. Run a new file over the steel. The softer the metal, the more easily the file will cut. (When the file slides easily over hardened steel, it is said to skate.)

Work in an area where you can control the amount of light. During annealing and hardening the light has to be dim; when tempering, you will need a very bright light. One solution is to do the first two steps inside, and the final step outside in the sun.

Ignite the MAPP gas burner and run a medium flame lightly over the blade, back and forth about an inch from the edge. Move the flame slowly until the metal turns a cherry red. Avoid letting it turn orange. (See Illus. 6.) Those woodworkers who find it difficult distinguishing between the different colors may prefer using an easier way to determine the proper temperature, which involves the use of a magnet. At the proper temperature, the steel will no longer be attracted to the magnet.

When the metal reaches a cherry color, quickly push it into the can of packed lime or ashes. These materials are good insulators and will allow the metal to hold its heat for a long time, letting it cool very slowly. It may be as long as 15 minutes before it will be cool enough so that it can be touched.

After annealing, the steel will be as soft as it can be. At this point, test it again with the file, contrasting its current hardness with the hardness it had before being annealed. Although this

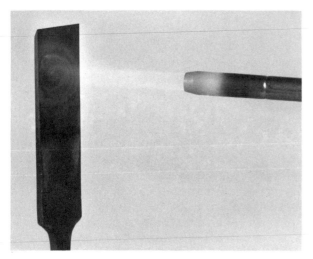

*Illus. 6. When annealing, heat the tool until it is cherry red and is no longer attracted to a magnet. Thrust it into a can of lime and leave it until it is cool.*

*Illus. 7. To harden a piece of annealed steel, heat it until it is cherry red and is no longer attracted to a magnet. Then plunge it into water; agitate it to prevent the creation of any air bubbles.*

step is not necessary, it will give you a feel for, and a better understanding of, the process.

If you have to reshape or regrind the edge, do most of it after annealing. If you do reshape the blade, do not try to file or grind the bezel to a sharp edge, as you would a piece of tempered steel. Leave an amount that's about the thickness of a pencil line.

When you have completed any grinding that's needed, reheat the area that has to be hardened. Do this in the same manner you used to anneal it. When the metal has again reached the cherry color, quickly plunge the cutter into the water. (See Illus. 7.) Agitate it so that bubbles of steam are not able to build up. Steam pockets will insulate the surface and interfere with the conduction of heat into the water. This may cause soft spots or, worse, it could even warp the blade.

After quenching, once again test the steel with a file. It is now fully hardened and is almost the same hardness as the file. The file will cut very lightly, if at all. If it still cuts the steel, check to see if the blade is made of wrought iron with a layer of steel welded to it (how and why such forge-welded layers were made, as well as how to detect them, is explained below and again in Chapter 9). If the blade is wrought iron with a layer of steel, you may not be filing steel, but rather iron, which cannot be hardened. Try the file on the opposite surface of the blade.

If the steel did not harden, add salt to the water to form a brine and try again. If this does not work, the steel is not tool-quality and cannot be satisfactorily hardened.

Before continuing, it is necessary to understand two terms: hardness and toughness. Heating the annealed steel until it is cherry red and then quenching it has made the steel as hard as this particular piece can be. If you were to sharpen the steel now, it will hold an edge better and longer than it will later, when you temper it. However, when hardened, the steel is also brittle and, if stressed, will easily break. Although hardened, the steel is said to lack toughness, which is the ability to withstand shock.

Steel is at its toughest when it is annealed, but in that state it cannot be sharpened. These problems are overcome by the third step in heat treating, called tempering.

Depending on the steel's purpose, tempering is the best possible compromise between hardness and toughness. Tempered steel is not as hard as it is possible to get, yet it is still hard enough to hold an edge. On the other hand, by making it less hard you will also make it tougher (less brittle) and more able to withstand the stresses of regular use.

Tempering can be done in two ways. Toolmakers usually use an oven because the temperature can be more closely controlled. An oven is also most efficient when a batch of blades are being tempered simultaneously.

You can temper steel in a household oven. There is a problem in that the thermostat is probably not very accurate. Therefore, test the thermostat's accuracy with an industrial mercury thermometer. Ignore the gradations on the oven's temperature dial and make your own mark at the point that gives you the desired oven temperature.

Another method is to use Temple sticks, which are available at industrial supply houses. These sticks look like crayons and are useful in tempering because they melt at a specific temperature. Rub the Temple stick on the metal and, like a crayon, it will leave a smear of color. When the metal is heated to the proper temperature, the smear will melt.

Tempering can also be done by color. (See Illus. 8.) Although this is less exact than using an oven, it is the easier method if you occasionally need to temper just one tool. It saves you the need to calibrate your oven or purchase Temple sticks.

Before tempering with the color method, use 220 grit sandpaper to polish the steel, which will be blackened after being heated with the MAPP gas. Next, using a buffing wheel (see Chapter 8) polish the area to be tempered to a mirror surface. The steel will change color as it is heated. These changes are subtle, and a polished surface will allow you to more easily distinguish them. Bright light is necessary to detect these changes in color, so you should now move outdoors.

Slowly heat the tool with the MAPP gas. Move the metal steadily through the flame, making contact about an inch behind the edge. Pause regularly, watching closely for the colors that reveal when the hardness is being drawn. There is a specific shade of color for about every 10 degrees of temperature. The first color to occur on the shiny steel will be a very pale yellow, indicating a temperature of about 430 degrees F.

As the steel becomes hotter, the intensity of the yellow will begin to increase. More heat will cause the color to change from yellow to brown, to purple, to blue. Exact colors and their equivalent temperatures are shown on page 24.

For most edge-tool cutters, the best compromise between hardness and toughness occurs at straw yellow, 460 degrees F. As the temperature climbs beyond this stage, the steel will become softer and softer and the edge will dull more and more quickly. Of course, the steel will also become tougher.

Tempering by color rather than in an oven has an added advantage besides convenience. As the tool is gradually heated, the heat flows forward from the point of contact with the flame towards the cutting edge. This means that the steel in contact with the flame is being made hotter than the cutting edge. It will become softer and, as a result, tougher than the edge. This is helpful on such tools as mortise chisels, which have to endure a lot of shock. Tempering with a flame results in an edge that is hard enough to remain sharp. However, the area of the blade above the edge is softer and tougher, right at the point where more toughness is most often needed.

*Illus. 8. When tempering by eye, be aware that different colors indicate different degrees of hardness. Because this photograph is black and white, the different colors appear as shades of grey. The dark blue on the steel (where it was heated) reads as a pale grey. The darker shades near the cutting edge are really straw yellow.*

Using a flame is fine for narrow edge tools such as chisels and plane blades. However, if you want to temper a tool that is much wider, such as a large plane iron or a drawknife, the flame is too concentrated to heat the entire width of the blade evenly. Let's review a technique that Peter Happny, one of the United States' best-known art blacksmiths, uses to temper a wider tool. Peter once heat-treated an adze head for me. A previous owner had put the tool in a fire to burn the end of a broken handle out of the eye. This ruined the temper.

First, Peter annealed the adze head and then hardened it. Both times he heated the entire tool in his forge. He then polished the area to be tempered. Next, he selected a block of scrap iron that was slightly wider than the blade. He heated this block in his forge until it was cherry red. With a pair of tongs he lifted the metal block out of the forge and placed it on a metal table.

Next, he held the adze head against the block, making contact about an inch behind the edge. The heat from the block was conducted into the head and gradually dissipated toward the edge, causing the colors to flow. When the edge reached a straw-yellow color, he quenched the adze. I used the tool for approximately a dozen years to rough-out Windsor chair seats.

After tempering, test the steel one more time with a file. You should be able to cut the steel with the file, yet it should be much harder than when it was annealed. Now finish grinding and, finally, hone the edge. Sharpening is the subject of Chapter 8.

Below is a list of temperatures and their corresponding colors:

| Degrees F. | Color of Steel |
| --- | --- |
| 430 | very pale yellow |
| 440 | light yellow |
| 450 | pale straw yellow |
| 460 | straw yellow |
| 470 | deep straw |
| 480 | dark yellow |
| 490 | yellow brown |
| 500 | brown yellow |
| 510 | spotted red brown |
| 520 | brown purple |
| 530 | light purple |
| 540 | full purple |
| 550 | dark purple |
| 560 | full blue |
| 570 | dark blue |
| 640 | light blue |

To get the most out of your tools, you sometimes have to make replacement cutters and even special cutters. In later chapters, I will explain how and why to do this for any particular tool. Tool steel is available in different widths and thicknesses and can be bought at most machinist supply houses. Ask for 01 steel. This is oil hardening steel, which can be quenched in motor oil rather than water. Otherwise, follow the heat-treating process just explained.

You will often find the words Cast Steel on many tools made before World War I. This term is somewhat confusing. It does not mean the blade was cast to shape. Cast steel (also called crucible steel) refers to a process for making steel rather than for making edge tools; in this process, the steel was made in a furnace and cast into ingots. Cast steel was highly prized by toolmakers. It is a water-quenching steel and is heat-treated as described above.

When heat-treating the cutters of older tools (wooden bench planes, moulding planes, wooden special-purpose planes, drawknives, and some chisels), you should be aware that until this century tool steel was costly. As a result, cutters were made less expensively; a thin layer of steel was forge-welded to the side opposite the bezel. The rest of the blade was made of cheaper wrought iron. This means that just the cutting edge is steel and capable of being hardened and sharpened (see above). If you look closely, you can often see the line between the two metals. This line is sometimes more obvious on an edge. (See Illus. 9.) The steel is also a lighter color than the wrought iron.

These blades have some advantages. As already explained, the hardness that allows steel to hold a sharp edge also makes the material brittle so that it breaks when stressed. Wrought iron (which cannot be hardened) is soft but tough. Because the thin layer of steel is backed up with wrought iron, it is less likely to break when the tool is used.

Blades with a forge-welded layer of steel also have a disadvantage. After many grindings, the steel can be completely ground away. Since this

*Illus. 9. The blades of earlier tools were often made of wrought iron, with a layer of steel forge welded to the upper surface. The line between the wrought-iron cutter and the layer of steel can be seen just in front of the hexagonal hole.*

can take a lifetime, it was not a problem for the tool's original owner. However, since some second-hand tools have been in use for a century or more, some have very little steel remaining.

Blades with a forge-welded layer of steel have distinct up-and-down sides. Reversing the blade is a mistake that would result in a cutting edge made of soft iron. Most often, the cutter already has a bezel ground on it and you do not have to worry about whether the blade is upside right or not. But as an added precaution, remember that the maker's stamp is usually on the up side.

# 7
# The Lapping Table

Before either a new or a second-hand tool can be used, it usually has to be sharpened and tuned. I will describe tuning individual tools in the chapters that deal with that type of tool. Sharpening is the subject of Chapter 8.

Much of the tuning and sharpening you will have to do requires a lapping table for flattening and removing rust. For example, you usually have to flatten a plane's sole before it will work well. Second-hand tools that have not been used for a long time are often rusted, and rust pits have to be removed from the cutter as the first step in sharpening. Also, on most edge tools (like a plane blade), you achieve a sharper edge if one side of

the cutter is perfectly flat. All these operations are done on a lapping table.

The lapping table is not really a table at all, but rather a perfectly flat surface. The table can be made of cast iron that has been machined flat; you can use, for example, an old planer or jointer bed. However, for several reasons I recommend a piece of plate glass as an alternative. The glass is easier to find, is more portable, and is easier to store.

A piece of plate glass can be purchased from a glazier. It should be at least 10 inches wide and *at least ¼ inch thick*. Regular window glass will break, and you risk injuring yourself. The length of the piece of glass depends on the length of the

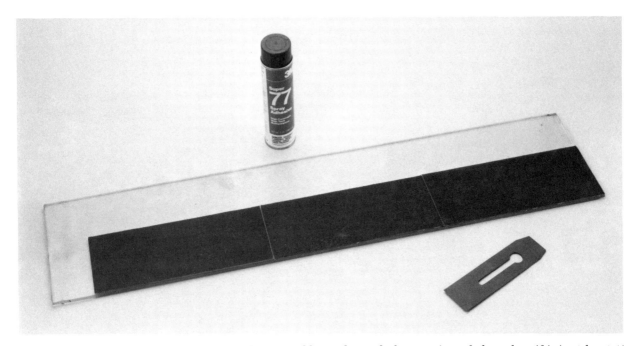

*Illus. 10. A lapping table can be made from a piece of plate glass if it is at least ¼ inch thick. If the table is long and narrow, bench planes can be lapped more easily. This photograph shows three half sheets of #220 grit emery cloth that have been butted together to produce a 30-inch surface. They were glued to the glass with a spray adhesive that's available from hardware dealers.*

tools you will be using on it. If you only use it for flattening cutters, you can use a piece that is as small as 10 inches square. For a long tool (like a jointer plane), your lapping table should be nearly twice as long (about 40 inches). (See Illus. 10.)

Attach a sheet (or sheets) of either emery, aluminum oxide, silicon carbide, or garnet paper to the surface of the table. The type of paper used depends on the amount of lapping you have to do. Emery cloth will hold up the longest, but it is the most expensive. It is good to use if you are tuning a number of tools at once. If you have just one job, use a sheet of garnet, silicon carbide, or aluminum oxide.

The fineness of grit used depends on the amount of metal to be removed and the speed at which you want to do it. If you have to remove a lot of metal due to heavy pitting, use 100 or even 80 grit sandpaper. These grits of sandpaper work very fast but also leave behind fairly heavy scratches that have to be removed with finer grits. For periodic flattening of chisels and plane blades, use 220 or even 320 grit sandpaper. Practice so that you will learn which grits to use for particular operations.

Creativity also plays a part here. If all you have to flatten are small surfaces such as chisels, you can adhere paper of several different grits to the lapping table. On the other hand, if you are jointing the sole of a bench plane, you need a long, narrow surface. Cut a 9 x 11-inch sheet across the width. Then, butt the two sheets together for a surface that is 18 inches long, but only 5½ inches wide (most planes are only 2–3 inches wide). If you are jointing the sole of a jointer plane (22–24 inches long), butt four half sheets together for a surface that is 36 inches long. Other ways to use the lapping table will be discussed in later chapters.

You can attach abrasive sheets with tape. However, paper with tape will often tear. It is better to use a spray adhesive that can be bought at a hardware shop. The adhesive bonds the entire sheet to the table. When the paper becomes worn, you can remove it by soaking it with mineral spirits. Then, scrape off the loosened paper with a single-edge razor.

When working on several projects at once or a single large job, keep a wisk broom handy and periodically sweep the paper free of the metal dust. If you do not, the paper will become clogged until it can no longer abrade the metal.

# 8
# Sharpening Techniques

I have been doing seminars and teaching wood-working classes throughout the United States for about eight years. In that time, I have met and talked with thousands of woodworkers, and have learned that the greatest impediment to using hand tools successfully is the average woodworker's inability to sharpen them.

Most woodworkers understand intuitively how any particular tool functions, but cannot make it work properly because he or she does not know how to sharpen it. If a tool is not razor-sharp, it will require much more effort to push or pull, and the cut will be ragged. As a result, too many woodworkers set their hand tools aside in disgust and turn instead to their machines.

Sharpening is not an arcane science. It is a skill that is very easy to learn. However, many wood-workers have been overwhelmed and confused by jargon and misconceptions. They needlessly worry that cutters have to be ground to an exact angle measured to the precise degree. Their anxieties are reflected in their most common questions: "What angle do you sharpen to?" "How often do you sharpen?" These questions will be answered in this chapter.

## Grinding the Cutting Edge

Before sharpening, begin by grinding the cutting edge to the desired shape. For example, if you are

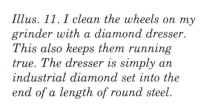

*Illus. 11. I clean the wheels on my grinder with a diamond dresser. This also keeps them running true. The dresser is simply an industrial diamond set into the end of a length of round steel.*

tuning a jointer plane which requires a straight-edge that is at a right angle to the sides of the blade, or a jack plane that has a curved cutting edge, do that first.

To determine the shape needed, first coat the side opposite the bezel with layout fluid. This is a fast-drying liquid that comes in several colors, although blue and red are the most common. Layout fluids can be either applied with a spray can or with a dauber. Once the fluid is dry, trace the edge's contour in it with a scratch awl. Even a very fine line is easily visible in the layout fluid. Finally, shape to that line. When finished, remove the layout fluid with a special solvent. Both layout fluid and the solvent can be purchased from industrial supply houses and are even carried by some larger hardware shops.

For grinding, I recommend a two-wheeled bench grinder. Grinders that use sanding belts make a bezel that is too long. They are also very difficult to hollow-grind with. The stone I use the most often on my grinder is 100 grit (fine). If you too decide to use a bench grinder, keep the wheels running true by regularly dressing them. I use a diamond dresser, which is a short metal rod with an industrial diamond set into one end.

It is available at industrial supply houses. The dresser also freshens the surface of the wheel and reveals new abrasive particles. (See Illus. 11.) By keeping the edge free from clogging and glazing, the wheel creates less heat; however, you should still keep a container of water by the grinder and cool the cutter often.

Whatever type of grinder you use, practice until you have the control necessary to create the cutting-edge shape your tools require and you are able to grind without causing so much heat that the cutting edge turns blue and becomes too soft to hold a sharp edge.

# General Sharpening Information

Following is sharpening information that applies to all tools. Any other special operations unique to a particular tool will be further explained in the chapter on that tool.

A sharp edge begins with two surfaces coming together at an angle. The sloping surface (the one that is ground) is called the bezel. Many woodworking magazines and authors also refer to it as the "bevel."

The bezel on many tools (planes, chisels, etc.) should be at about 30 degrees. There are some exceptions, and these are noted in the appropriate chapters. Don't be dismayed by the reference to angle degrees. You only have to approximate the angle. It is more important that you learn to sharpen so that you can use and enjoy your tools.

I prefer to hollow-grind my tools, which means that rather than being a true, flat surface, the bezel is slightly concave. If you are using a bench grinder, this shape will occur naturally. A hollow grind creates an edge and a heel. (See Illus. 12.)

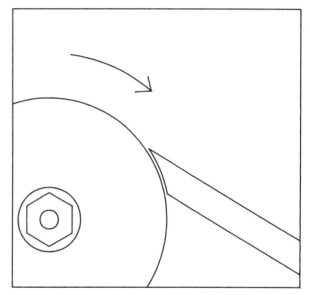

*Illus. 12. The round edge of the grinding wheel creates a concave bezel which is said to be hollow ground.*

When sharpening a tool, both the bezel and the opposite side have to be flattened and polished, in a process called honing. I instruct students to bring their tools to class already sharpened and tuned. When a student's tool does not work, I disassemble it and usually find that he or she has spent a great deal of effort honing the bezel, but has completely ignored the opposite surface. (See Illus. 13.) So, do not even touch the bezel until you have first flattened this side. Unless this surface is flat, the tool will never perform to its potential. Depending on this surface's condition, it is possible that you need to remove a lot of metal, which

*Illus. 13. A new replacement plane iron purchased from a hardware dealer. The scratches created by the surface grinding done at the factory result in a uniform matte sheen. Flatten and polish the front edge until you can see your eyes reflecting in it. Otherwise, it will not be sharp enough to cut well.*

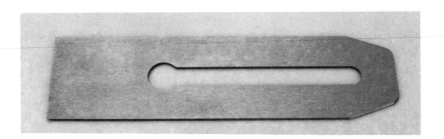

is why you should first flatten the side opposite the bezel on the lapping table when sharpening either a brand-new or second-hand tool or one that you have not maintained.

If the surface does not have obvious pitting or depressions, use 220-grit paper. Attach it to the surface of the table. Lay the blade flat on the lapping table. It must be kept perfectly flat; at no time should you raise it.

After the first dozen or so strokes, stop and look at the back of the blade. (See Illus. 14.) The high

*Illus. 15. Once the entire cutting edge has a uniform sheen, polish the side opposite the bezel with a honing stone. You'll know you have polished this side correctly if you can hold the blade up to your eyes and see your reflection.*

*Illus. 14. Note the high spots (the shiny areas) on this bench plane blade, which has spent a short time on the lapping table. The blade has to be lapped until the high spots are levelled and there is a uniform sheen along the entire cutting edge.*

points have been in contact with the paper and will be lighter in color than the low spots. This will allow you to gauge how many more strokes the blade needs. The job is done when the entire surface has a uniform sheen. (See Illus. 15.)

The lapping table flattens the back of the tool

quickly, but it also produces a mesh of fine scratches. This surface is too rough to result in a sharp edge and needs more work. Now this surface should be polished. Polishing is done on whetstones. You should have at least a medium- and a fine-grade stone. It really does not matter what type you use; just make sure that the surface of the stone is itself flat.

Japanese waterstones are currently popular with woodworkers. They are inexpensive and come in a variety of grits. However, they have to be kept wet, which means that you have to store them in a tub of water. Since the water can splash, you also have to create a separate sharpening station. Waterstones are soft and wear quickly, but can themselves be flattened on the lapping table. Use 440 or 320 wet or dry paper (silicon carbide).

Recently, I have begun using man-made ceramic stones, which are very hard and seem to never lose their flatness. They are used dry, which I find more convenient, especially when travelling. When their surfaces become glazed with metal dust, clean them with a sponge and bathroom scouring powder.

Oilstones, both natural and man-made, are the more traditional alternative. They tend to be more expensive, but are usually more durable than waterstones. They have to be lubricated with oil, and leave a dirty oil film on your fingers. This film gets on everything you touch.

When polishing the side opposite the bezel, hold the blade perfectly flat, like you did on the lapping table. Once again, periodic inspections are necessary to determine how well the polish is developing. The entire surface does not have to be polished, only the area along the cutting edge. Look very closely to be sure that the polish extends right out to the very edge. I usually start with a medium stone, followed by a fine stone. The result should be a mirror polish, one that is so smooth you can see yourself in its reflection. Anything else will not produce a razor-like sharpness.

Now you can flatten and polish the bezel. It is important here that the tool does not rock on the bezel. If you hold it too low, you will be honing the heel. If you hold the tool too high, you will round the cutting edge and the tool will not become sharp. (See Illus. 16.)

Companies that sell new tools stock a device that holds the blade at a set angle. These devices are sometimes referred to as roller skates because they ride on a wheel. They are more complicated to use than they are worth.

I use a simpler method to keep the tool from rocking. Remember, I recommend hollow-grinding, so the bezel has a distinct edge and heel. First, set the heel of the bezel on the stone. This angle is, of course, too low to hone. Slowly raise the blade until the cutting edge is also touching the stone's surface. The bezel is now in contact at two points, the heel and the edge. If you hold it in this position while honing, you will have no problem with the edge-rounding.

Work the tool in whichever pattern you prefer—figure eights, circles, back-and-forth, or side-to-side. Take the tool off the stone after approximately a dozen passes; you should see a narrow line of polish on the heel and at the cutting edge. (See Illus. 17.) As long as both edges are being polished uniformly, the edge is not rounding. Because you are only honing these two narrow surfaces, honing a hollow-ground bezel is much quicker.

By honing, you have made a very narrow flat surface at the very outer edge of the hollow-ground bezel. This polished flat edge, which meets at an angle with the perfectly flat and polished opposite side, creates the condition called "sharp." When your edge tools are sharpened this way and tuned, they will perform to their utmost capacities.

If the tool is too long (or too big) to run over the stone, hold the stone and run it over the edge. (See Illus. 18.) Be careful to first place the stone in contact with the heel, and then roll it until it also touches the cutting edge.

Check the polish along the bezel. It should extend along the entire length of the edge and be of mirror quality.

After you have honed the bezel, the edge should be sharp. Test it with the tip of your thumb. If

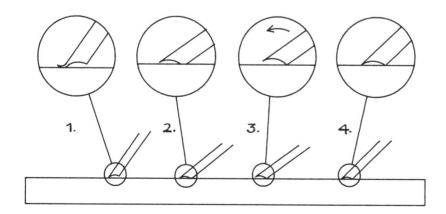

1.    2.    3.    4.

*Illus. 16. If you hold the tool too high when honing, you will roll the edge as shown in 1. If the tool is too low, the honing only takes place on the heel, as shown in 2. To locate the correct angle, place the heel on the stone as shown in 3 and raise the back end until the edge comes in contact with the stone, as shown in 4. Make sure that both surfaces stay in contact with the stone while you are honing.*

Illus. 17. Turn the blade over and hone the hollow-ground bezel; make sure that you ride the heel and edge on the stone. The result will be two fine lines of polish, which can be seen on this chisel. Some more honing is still necessary to make the cutting edge gleam with mirror-like polish.

Illus. 18. Some tools are too large to hone on a stone. Here I am honing a very large two-inch-thick chisel that I use in timber framing by holding the stone and running it over the edge.

your skin is snagged by a slight wire edge, strop it away. I do this on a buffing wheel impregnated with an abrasive compound. I use grey steel, but jeweler's rouge will also work. (See Illus. 19.) Buffing should amount to no more than a momentary touch against the wheel. If you take any longer, you risk rounding the edge and losing the sharpness you just created.

Next, with a bright light shining over your shoulder, hold the blade so that you are looking directly at the edge. If the edge is sharp, it will not be visible in this inspection. If you see even the narrowest sliver of surface glinting in the light, your tool is not sharp in that location.

Next, test the edge by shaving the end grain of a block of hardwood. (See Illus. 20.) The shaving should pare evenly and cleanly. Some woodworkers test for sharpness by shaving a hair on their arms. I lost interest in this method the day I watched someone shave off a patch of skin.

When you understand that a sharp edge is the meeting of two flat, polished surfaces coming together at an angle, you understand why blemishes such as pitting (or on a new tool, the tracks of the factory's surface grinder) can pose problems. If the bezel is polished, but the opposite side is left scarred, the edge is really a microscopic saw. You will exert more effort to move the tool, and the wood's surface will be scratched and scarred.

As you use a tool, the cutting action creates friction which, in time, blunts the edge and rounds the side opposite the bezel. Thus, the answer to the question "How often do you sharpen?" depends on how much you have used the tool, not the amount of elapsed time. It is time to resharpen the tool (usually by rehoning) when the tool no longer cuts easily and cleanly.

When resharpening the tool, your first step should be to hone the side opposite the bezel to reflatten it. If you hollow-grind, the polished surface along the edge will become wider with each honing. (See Illus. 21 and 22.) So will the one on the heel. Eventually, the two surfaces will merge so that you are honing the entire bezel. This takes a lot of time and effort and also creates more wear on your stones.

At this point, it is time to grind again. Each time I grind the bezel I also return the cutter to the lapping table and completely reflatten the opposite side.

*Illus. 19 (above left). Impregnate the surface of your buffing wheel with a buffing compound. In this case, I am using grey steel that is suspended in a wax block. Friction creates heat, which melts the wax, and as a result the compound is deposited on the pad's surface. Illus. 20 (above right). Test the sharpness of a cutter's cutting edge on the end grain of a piece of hardwood. It should cut a clean shaving with little effort.*

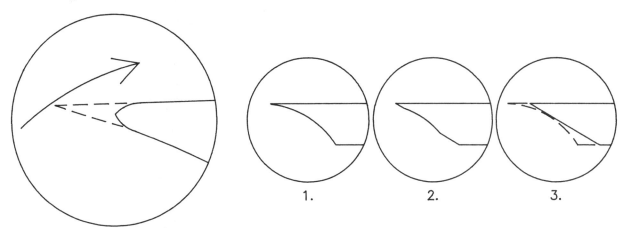

*Illus. 21 (above left). Cutting action and the passage of the chip over the edge create friction that wears both the bezel and the opposite side. This is one reason why the first step in sharpening and honing is the flattening of the side opposite the bezel. Illus. 22 (above right). 1 shows a hollow-ground edge. Honing creates narrow, flat surfaces on the heel and edge, as shown in 2. Each honing removes metal from these surfaces, increasing their widths. Eventually, these two surfaces will merge, as shown in 3. When this happens, it is time to regrind. If you do not, and instead continue to hone, you have to remove metal from the entire width of the bezel. This is a lot more work, is very time-consuming, and is hard on honing stones.*

Hollow-grinding is fine for bench planes, some special-purpose planes, and paring chisels. However, the hollow bezel makes edges thinner, and, thus, more fragile. I do not hollow-grind tools that have to withstand a lot of shock, or that do heavy work like, for example, the curved blade of a jack plane, mortise chisels, and plow irons. Moulding plane irons are shaped with files (see pages 34–36) and should not be hollow-ground either.

# Slipstones

Whetstones are fine for most bench plane irons and chisels. However, some cutting edges (for example, moulding plane cutters) have a complex profile; to hone this type of edge, you need specially shaped stones. Most slipstones have a cross section like an elongated teardrop. This results in two flat sides with two round edges, one wide, the other narrow. The ends have square corners. (See Illus. 23.)

Slipstones are made of the same natural and man-made materials as whetstones. The Japanese make water-lubricated slipstones. All slipstones come in the same grades as whetstones.

Long, narrow stones with special shapes are also available and are handy for many sharpening jobs. These stones are called files (sometimes "gun files"). This can cause confusion, since metal files (see following head) are also used in sharpening. Following are some of the cross sections files have: triangular, round, teardrop, square, and knife edge. Files are usually made of man-made materials. I use a set of ceramic files. (See Illus. 24.)

# Files

Many tools (for example complex moulding plane cutters) cannot be sharpened on a grinding wheel. Some woodworkers use small hand-held grinders,

*Illus. 23. A group of slipstones. The different colors represent different degrees of fineness. Note that the slipstones taper in thickness and have round edges that make them handy for honing gouges and moulding plane cutters.*

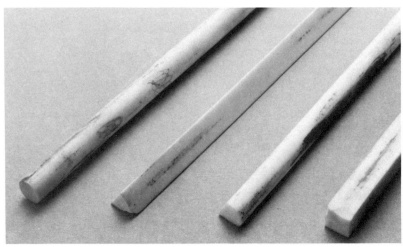

*Illus. 24. A group of ceramic files with various cross sections. From left to right, they are as follow: a round file, a triangular file, a tear drop file, and a square file.*

the type used in power-carving. I prefer metal files. They are less expensive and you have more control.

Most woodworkers know very little about files. These tools are used in numerous trades and come in an amazing variety of sizes and shapes. Most are useless for tuning and sharpening second-hand woodworking tools. As a result, the information in this section is limited to those files that can be used. (See Illus. 25 and 26.)

A file removes metal because of the cuts (called teeth) in its surface. Files are made in varying degrees of fineness and coarseness that are determined by the number of teeth per inch.

A bastard cut file has the fewest number of teeth, which means the cut is relatively coarse. A second cut file has more teeth, so it is smoother. smooth cut files cut, as they name suggests, smoothly. "Dead smooth cut" files have the most teeth per inch, but these files are not easy to find.

If a file has rows of parallel teeth (usually at an angle to the edges), it is called a single-cut file. Single-cut files cut with light pressure and produce a smooth surface. If two rows of teeth overlap each other to produce a cross-hatched surface, the file is a double-cut file. Double-cut files are used with heavy pressure and cut quickly, but less smoothly.

## Mill files

08354

For sharpening mill or circular saws. Also for draw-filing and finishing metals. All sizes tapered slightly in width. Two square edges. Single-cut on sides and edges.

| |←L→| "/mm | Catalog & Nida/Sida Numbers Bastard | Second Cut | Smooth | Wt. per doz. lb., oz/g | Shelf Pack |
|---|---|---|---|---|---|
| 4/113 | 08243 | – | 08306 | 0,08/227 | 12 |
| 6/150 | 08354 | 08385 | 08416 | 1,03/539 | 12 |
| 8/200 | 08497 | 08529 | 08560 | 2,12/1247 | 12 |
| 10/250 | 08642 | 08673 | 08704 | 5,07/2466 | 12 |
| 12/303 | 08737 | 08768 | 08799 | 9,01/4111 | 6 |
| 14/356 | 08832 | – | 08894 | 14,00/6350 | 6 |
| 16/400 | 08925 | – | – | 21,00/9526 | 6 |

## Mill files – 1 and 2 round edges

09099

Same as regular mill files except that they have one or two round edges. Bastard cut, single-cut on sides and edges. Round edges are used where rounding gullets are preferred, as compared to sharp corners or squared gullets.

**One round edge bastard cut**

| |←L→| "/mm | Cat./N/S No./No. | Wt. per doz. lb., oz/g | Shelf Pack |
|---|---|---|---|
| 8/200 | 09099 | 2,13/1276 | 12 |
| 10/250 | 09179 | 5,01/2296 | 12 |
| 12/303 | 09227 | 9,01/4111 | 6 |

**One round edge blunt 2nd cut**

| |←L→| "/mm | Cat./N/S No./No. | Wt. per doz. lb., oz/g | Shelf Pack |
|---|---|---|---|
| 12/303 | 10260 | 9,01/4111 | 6 |

**Two round edge bastard cut**

| |←L→| "/mm | Cat./N/S No./No. | Wt. per doz. lb., oz/g | Shelf Pack |
|---|---|---|---|
| 8/200 | 09389 | 2,11/1219 | 12 |
| 10/280 | 09469 | 5,01/2296 | 12 |

*Illus. 25 and 26. These two illustrations, reprinted from a Nicholson catalogue, depict mill files, common files that can be used to tune and sharpen tools.*

The end of a file you hold has a tang; the tang's shoulders are called the heel. The other end is the point. The distance between the heel and point is the length of the file.

Files vary in shape; some are flat. Their wide surfaces are called the sides. The narrow surfaces are the edges. If an edge or side does not have teeth, it is said to be safe. When the edges are parallel, the file is blunt; otherwise, it is tapered. Some files taper in width. Others taper in thickness; some taper in both dimensions.

The most common file is the mill file. These files taper slightly in width. Both the sides and edges are single cut. Flat files taper in both width and thickness. They have double-cut sides and single-cut edges. Warding files are very thin, with an extreme taper towards the point. They have double-cut sides and single-cut edges.

Files come in other shapes that are handy when sharpening or tuning tools. The cross section of a three-square file is an equilateral triangle. These files can either be tapered or blunt. Round files are also available as either tapered or blunt files, although I usually purchase chain-saw files. These files are sold in seven diameters from $\frac{1}{8}$ to $\frac{3}{8}$ inch. Needle files are very small (about $5\frac{1}{2}$ inches long) and are useful for very detailed work. They are usually sold in packages of six different shapes.

Most of the files described can be found in any good hardware shop. Industrial supply houses carry an even broader selection.

To use a file, lift it at the end of each stroke; do not drag it back to the starting position. When filing a thick edge or a flat surface, hold the file at a slight angle. This is called draw-filing, and it will cut metal very quickly.

If you do a lot of filing, get a file card. This is a special brush with widely spaced, short metal bristles. When it is pulled over a file, it will remove metal filings that are clogging the teeth.

If cleaned regularly, your files will last longer. Always store files so that they are protected and cannot rub against each other or other metal objects.

# Bench Planes

# 9
# Wooden Bench Planes

Woodworkers have almost always relied on a group of tools called bench planes to turn rough-sawn lumber into finished boards. These tools are called bench planes because they are generally used to work a piece of wood that is either clamped to the top of a workbench or held in one of the bench's vises.

Bench planes were made in different sizes (and sometimes different shapes) that corresponded to their various functions. They were used to smooth, square, and shape the parts that became furniture, architectural elements, and other wooden products.

Many modern woodworkers use the planer, jointer, and belt sander instead of bench planes. However, bench planes are still essential in shops that produce fine work. In fact, one indication of a woodworker's skill and the level of quality produced in his shop is the number of hand planes he uses and their condition.

Before the Industrial Revolution, bench planes were usually made of wood. In England and America the earliest planes were sometimes made by the craftsman himself. During the 18th century, plane-making became the specialty of individual toolmakers working by themselves or with a small group of apprentices and journeymen in what is known as the craft-shop tradition.

Craft-shop plane makers often used individual techniques and patterns. Today, knowledgeable collectors of these early tools can sometimes identify a plane's maker just by these features. Since these early planes (American and English) are not common and can be very valuable, I do not recommend trying to use them.

Interestingly, the plane-maker's craft is being revived by some modern woodworkers who make copies of early planes. You should be aware of these craftsmen for two reasons. First, if you cannot find a particular plane that you want to use, you can have one made. Second, the planes made by these modern craftsmen will eventually find their way to the second-hand market. Since they are handmade, they may appear to be older than they really are, and you may end up paying more for the plane than it is really worth.

During the mid 19th century, the making of wooden planes, like many other crafts, moved from the workshop into the factory. Nineteenth-century factory-made wooden planes were produced in very large quantities by tool companies operating in both England and America (mostly east of the Mississippi River). These factories exported their planes all over the world. Wooden planes continued to be made in large quantities until about World War I. In contrast to the highly individualized planes made by earlier plane makers, there was a surprising standardization among factory-made planes. As a result, they all look very similar.

English planes have always been in use in America. Throughout the 19th century, while America's domestic wooden plane industry was flourishing, English planes continued to be imported just as they had been during the craft-shop period. These English planes were also standardized and look very much like their American counterparts.

In England, handcraft traditions survived long after they had disappeared in America. As a result, wooden planes, as well as many other hand woodworking tools, were made and used in England after they had ceased to be made and used in the United States.

For the past two decades, American tool dealers have been scouring England for wooden planes. As a result, many of the planes sold today in American shops are English. English planes are

*Illus. 27. An 18th-century handmade jack plane (top) with a single iron and a 19th-century factory-made jack plane with a double iron.*

generally less valuable to American collectors, so they tend to cost less. For woodworkers who want to use tools rather than collect them, they are often good bargains. These planes can be of very good quality.

The supply of factory-made wooden bench planes (English and American) is so large that these tools are commonly available everywhere. Any woodworker who wants to use them will have no trouble finding good, serviceable tools.

# Maker's Marks

Because factory-made planes (both of American and English origin) look so much alike, they cannot generally be identified and assigned to an individual factory by distinguishing characteristics. However, most plane-making companies marked their products by stamping their name on the end grain of either the heel or toe. (See Illus. 28.) This type of identification was so common that it is unusual to find a factory-made plane that is not stamped. If you do acquire a plane that is stamped with a maker's mark, you can look it up in the complete checklists of both English and American plane makers that can be found in W. L. Goodman's *The History of Woodworking Tools* and Emil and Martyl Pollack's A *Guide to American Wooden Planes & Their Makers.* (See the Bibliography on page 254.)

The maker's stamp is not the only mark found on wooden planes. The craftsmen who owned these tools commonly added their own names or initials. Over many decades, wooden planes were owned by several generations of woodworkers. When a plane was eventually passed on to another man, he, too, usually applied his stamp. The second or third mark was often intentionally

*Illus. 28. The toe of this smooth plane bears several stamps. The first is that of M. Wilson, 75 Clay St., New Castle (England), who may have been either the maker or a merchant. The plane was owned by G. Strong, who used his stamp to eradicate that of an earlier owner, D. Cutherburtson. At the top is the word "Cabinet," which may be a brand name.*

applied over previous marks to obliterate them. The new marks were then put on (more clearly) somewhere else on the tool.

I have seen wooden planes whose ends are a mishmash of owners' stamps; these stamps crowd and overlap each other so much it is nearly impossible to make out any of the names. I own one plane where a second owner even went so far as to drill out the end grain where another man's initials were. He filled the holes with mahogany dowels and stamped his own initials on the end grain of these plugs.

Of the many marks stamped on a plane, one can be that of the maker, another the merchant who sold the tool and who applied his own mark, and others of one or more owners. Usually a plane maker or a tool dealer would use a more elaborate stamp than would a woodworker. This is sometimes the key to identifying the marks.

# Wood Used

Most American and English wooden planes are made of beech. This wood is highly shock resistant, which is an important quality because of the way in which the iron is removed and adjusted. Beech is difficult to work because of its extreme hardness, but it takes a high polish when used. A wooden plane that has been cared for has a glassy surface that I find very pleasant.

Nineteenth-century plane makers often advertised that they used only the best-quality second-growth beech to make their plane bodies. Some plane makers claimed that they used wood that was split from the log, instead of plank bought at a sawmill. Since splitting follows the grain, the plane maker who used wood that was split from a log was assured that the wood he was using was free of any defects such as encased knots or a wavy figure.

By splitting their wood, these factories could also obtain a radial section (as opposed to a tangential section). Imagine that the end grain of

a beech log is a pie. A radial section would be a slice of the pie. Thus, splitting the wood resulted in plane bodies that were less inclined to warp or change dimensions due to seasonal movement.

In America, planes were also made of yellow birch, hard maple, and hornbeam. Other, less common, native woods like live oak were occasionally used. Live oak is an American tree that grows in the southeastern coastal plains. Its grain is very dense and looks somewhat like burl. A live oak plane is much more resistant to wear than even beech.

Some imported species of wood made even better plane bodies than did these native varieties. For example, lignum vitae is a heavy, hard, waxy wood that is so dense it will not float. A partial list of other imported woods that were occasionally used to make bench planes would include satinwood, boxwood, ebony, and rosewood.

# Plane Parts and Characteristics

All bench planes have certain parts in common, and their names reflect a certain anthropomorphism. If you have a wooden plane, hold it in your hands and look at these various parts. (See Illus. 29.)

The main part of the tool is its body (sometimes called the stock). The body has a heel at the rear and a toe at the front. The bottom of the body is called the sole. The opening in the sole through which the cutting edge of the iron projects, and through which the chip passes, is called the mouth. On a wooden plane, the chip passes through the mouth and up into a cavity called the throat. The sides of the throat are called the cheeks.

*Illus. 29. A cutaway view of a wooden jack plane. Most bench planes have the same features. Most smooth planes, however, do not have a tote.*

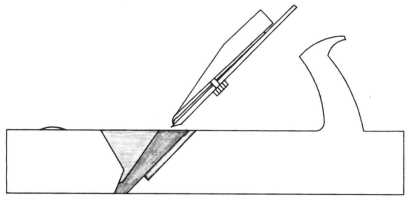

If you remove the wedge, you can see more of the plane's anatomy. The tapered slots in the cheeks that hold the wedge and iron in place are called the abutments. The front end of the throat is a surface (inclined towards the toe) called the wear. The iron rests on a surface that is inclined towards the plane's heel, and which is called the bed. Many of these terms were later passed on to metal planes.

## Totes

Most wooden bench planes (with the exception of the coffin-shaped smooth plane) have a handle that is gripped with the right hand. This handle is called the tote. Totes come in two varieties. (See Illus. 30.) A closed tote is similar to an inverted saw handle, while an open tote looks like an inverted pistol grip. An open tote is most commonly found on modern metal planes.

### RAZEE PLANES

The stocks of most bench planes are of a single thickness from heel to toe. However, on some planes the stock narrows in thickness just behind the iron. These are called razee planes. (The word may be a corruption of the past tense of the French word *razer,* which means to shave.) In

other words, the stock near the heel of a razee plane has been shaved away. Razee planes almost always have a closed tote.

Tool collectors often claim that the razee plane was developed for use in boatbuilding. However, I have spoken with many boatbuilders and none could explain why this type of stock would be better adapted for their trade.

Razee planes are more efficient than straight planes, and this is an advantage that all woodworkers will appreciate. The shaved area allows the tote to be set lower, so that the thrust applied by your right hand and arm is squarely behind the stock. Otherwise, the thrust is above the body of the plane. (See Illus. 31.)

Razee fore planes are common, and many are made of exotic woods. I have seen some razee fore planes that were made of lignum vitae and ebony. I own razee jack and try planes made of live oak, but do not recall ever seeing a razee jointer. A type of razee smooth plane was made by many plane factories. (See Illus. 32.)

## Pitch

The pitch, or bedding angle, is the angle at which the plane's iron is set. On most bench planes, this is 45 degrees, and is known as the common pitch.

*Illus. 30. A plane with a closed tote, and one with an open tote.*

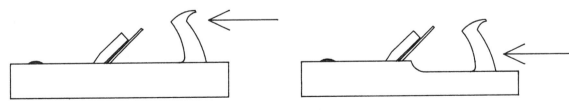

*Illus. 31. You can use a razee plane more effectively when you push it because the thrust of your arm is behind the stock rather than above it.*

*Illus. 32. A group of razee planes. The jack and fore plane in the foreground are the type that is often referred to as shipbuilder's planes. Both are made of live oak, a species of wood that was once commonly used in shipbuilding, but neither has a maker's stamp. The smooth plane (front left) was made by Auburn Tool Co. The fore plane (rear) was made by the Ohio Tool Co., and the jack plane by Scioto Works, a division of Ohio Tool.*

Such planes were meant for everyday use, and they work best on soft woods.

However, some planes were made specifically for working hard woods and have a pitch of 50 degrees, called a York pitch. Cabinetmakers who regularly worked heavily figured woods such as walnut or mahogany might own a smoothing plane with its blade set at an even steeper angle of 55 degrees. This is called the middle pitch. There is a more extreme set of 60 degrees known as the half pitch, also intended for heavily figured woods such as curly maple. (See Illus. 33.)

Planes with the middle and half pitch work like scrapers and are difficult to use on soft wood such as pine. Anyone who has ever tried to run a scraper over soft wood knows that the edge tends to pull up loose fibres rather than shavings.

# Irons

Modern irons are stamped out of sheets of tool steel and are of a uniform thickness. Wooden plane irons were usually tapered, being thicker at the lower end than at the top. (See Illus. 34.) This extra thickness strengthens the edge and makes it less prone to vibration when cutting. This vibration is called chatter, and it results in series of fine lines called chatter marks that mar the planed surface. Tapered irons have another purpose that will be discussed later on.

Curiously, tapered plane irons are still being made in mainland China, maybe because the country was "closed off" to Western influence for several decades. At least one American tool com-

pany has imported them, and at one time had them for sale in its catalogue. I do not know if they are still available.

Besides being tapered, wooden plane irons are different in another way. They are usually made of wrought iron with a thin layer of steel forge-welded to the forward, upper surface (see Chapter 6). If you look closely, you can often see the line between the two metals. The steel is also a lighter color than the wrought iron.

Many bench plane irons have been ground until little steel is left. In this case, you have a good reason for asking the dealer to lower the plane's price. (See Illus. 35.)

## MAKERS' MARKS AND ENGLISH IRONS

Plane blades were made by companies that specialized in the production of edge tools (chisels, etc.). These companies would usually stamp their own marks onto the iron's upper end, so that the mark is usually visible just above the wedge. (See Illus. 36.)

Some American factories made plane blades, but American woodworkers preferred English irons. This explains why it is common to find American planes with English cutters. In my observation, the reverse is not true. However, blades and chip breakers made in either country are often interchangeable. Lists have been compiled of marks used by these edge-tool makers, and they can help you determine whether an iron is English or American. A good, clear maker's mark increases a tool's value. An unusual or a

*Illus. 33. These smooth planes have three different pitches. From right to left, the pitches are as follow: half, middle, and common pitch. The first two planes are used on hard woods that have a prominent figure. A 50-degree pitch (called the York pitch) is not shown, but is also found on second-hand planes.*

*Illus. 34. A tapered iron from a wooden bench plane and a parallel iron from an iron plane. The thicker end of a tapered cutter helps the iron resist chatter.*

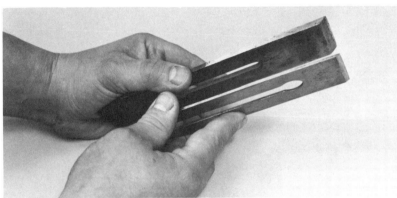

*Illus. 35. This cutter is worn out. It has been ground to the slot, and little or no steel remains. The price of a plane with such a cutter should be discounted.*

*Illus. 36. Bench plane cutters that bear maker's marks. From left to right, the maker's marks are: Hearnshaw Bros., Sheffield (England), with a John Bull trademark (complete with a figure of a rotund Mr. Bull); W. Butcher (single iron); Buck Brothers (Millbury, Mass.); and Moulson Brothers. A good, clear stamp on the iron increases the plane's value.*

rare mark is more valuable than a common one.

You will also often find the words Cast Steel stamped on many edge tools made before World War I. As explained in Chapter 6, this does not mean that the blade was cast to shape. The steel was made in a furnace and cast into ingots.

I purchase any loose plane irons that I find. When I find a bench plane I want to use, but does not have its iron, I always have a spare that's the exact size.

## Chip Breaker

The earliest wooden bench planes had a single tapered iron held in place by a wooden wedge.

This is called a single iron. The double iron (with a chip breaker) became common during the 19th century. (See Illus. 37.) The chip breaker is attached to the iron by a screw; the screw's head is slipped through a slot in the iron and tightened. The chip breaker, as its name implies, breaks the chip. As the chip rises through the mouth into the throat, it is pressed against the chip breaker, causing a continuous series of kinks that can be seen if the shaving is held up to the light. (See Illus. 38.)

If the chip were not broken, it would act as a lever. Lifted by the incline of the iron, the unbroken chip would, rather than being shaved off

*Illus. 37. At center is a double iron. Its chip breaker is on the right, and a single iron is on the left.*

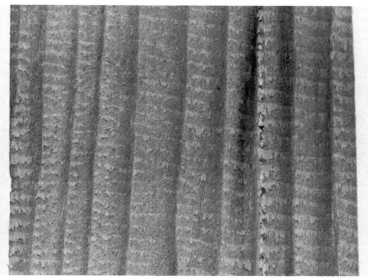

*Illus. 38. If you hold a plane's shaving up to the light, you will see faint horizontal lines. These are called breaks; they are kinks made in the chip to prevent it from tearing ahead of the cutting edge. The vertical lines are the wood's annual growth, often called its grain.*

cleanly, tear loose ahead of the cutting edge. (See Illus. 39.)

The chip breaker has another purpose. When you tighten the screw, you create pressure that bows the iron. This stiffens the blade and helps resist chattering. (See Illus. 40.)

A screw with a large head holds the chip breaker and the iron together. Often, the hole in the chip breaker is threaded. Other times, a separate nut is attached to its up side.

Thus, a bench plane with a double iron will have a slot routed in its bed to accommodate the screw's wide head. (See Illus. 41.) If there is a nut, it will be accommodated by a second slot routed in the underside of the wedge.

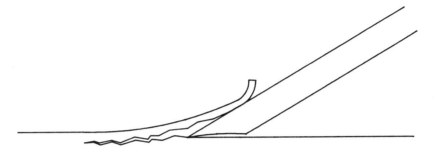

*Illus. 39. Without a wear and chip breaker to kink the chip, the cutter will act as a lever and tear the chip loose ahead of it.*

*Illus. 40. The chip breaker not only breaks the chip; when tightened, it also helps stiffen the cutter so that it better resists chatter.*

*Illus. 41. A bench plane with a double iron will have a slot cut into the bed to accommodate the head on the chip breaker's screw. If a nut is attached to the cap, a slot will also be cut into the back of the wedge.*

# Wedge

Wooden plane irons are usually held in place by a wooden wedge that has a narrow end that slides down over the chip breaker. (See Illus. 42.) Remember, the iron tapers in the opposite direction of the wedge's taper. The two tapers pushing against each other results in a tighter lock. (See Illus. 43.)

A wedge has two legs on its lower end. When put into the plane, the legs fit under the abutments. (See Illus. 44.) The ends of the legs are bevelled to form points, so as to not snag the chip as it passes up the throat. There is an opening between the legs, and above this space is another taper. This second taper helps direct the shaving up and out of the throat. (See Illus. 45.)

*Illus. 42. A wooden bench plane's cutter is held in place by a wooden wedge. Note that the wedge is shaped to help guide the chip up the throat. The legs are tapered in two directions to form sharp points. The space between the legs is also bevelled.*

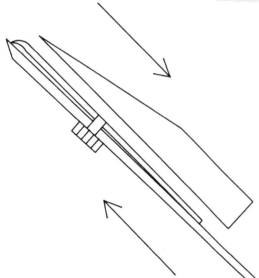

*Illus. 43 (above left). When the tapered iron and chip breaker are assembled, they produce a wedge. This assembly is tapered in the opposite direction from the wooden wedge. The result is two tapers that work against each other to produce a more secure lock. Illus 44 (above right). A wedge should fit tightly under the abutments, producing no gap in which the chip can become caught. Note how the bevel between the legs helps create a passage for the chip.*

Illus. 45. If you look down the throat of a bench plane with its cutter removed, you will see the tapered abutments that hold the wedge in place. The front surface is the wear, and the rear the bed.

Illus. 46. The cutter (single type) from a jack plane. Note that the cutting edge is curved to about an 8-inch radius.

Wooden planes from Japan are currently popular with woodworkers, but these planes do not use a wedge. Their irons are tapered, but in the other direction, so they are thicker at the upper end. The iron's wedge shape is all that holds it in the plane. If the plane's dimensions change due to seasonal movement (or if the abutments become worn), the iron cannot lock without projecting so far out from the mouth that the plane "chokes." It is necessary to place paper shims between the iron and the abutments to correct the fit. In this respect, American and English second-hand wooden planes are far superior.

# Types of Planes

Without their bench planes, preindustrial woodworkers would have been helpless. Their lumber came from the mill in an unfinished state, called rough-sawn. The gashes left by the saw were quite deep and before a piece of wood could be worked it had to be surfaced. It was a waste of time to use a plane that only removed a thin shaving. Obviously, something more effective was needed. The bench planes met this requirement. Following is a description of the different types of bench planes.

Illus. 47. Because the edge of a jack plane iron is curved, the plane can cut a heavy chip without chocking.

# Jack Plane

The first plane used in smoothing a rough-sawn board was called a fore plane by early cabinet-makers, and a jack plane by carpenters. By the mid-19th century, the name jack plane became universal and the term fore plane was used for another bench plane that had previously been called a try plane.

A jack plane's body is about 15 inches long. But the feature that most distinguishes this tool is its cutting edge, which is ground to a distinct convex curve with a 5–8-inch radius and allows the plane to cut a shaving as much as $\frac{1}{16}$ inch thick. (See Illus. 46 and 47.) If the edge were ground square, it could not take a heavy chip because the corners would dig into the wood and choke the plane. To take a look at the cutting edge, just turn the tool over or remove the iron.

# Fore Plane

This tool, also used in making stock from rough lumber, was until the mid-19th century called a try plane because a woodworker was said to "try" the surface of a board when he used this plane. (See Illus. 48.) After the mid-19th century it became known as a fore plane. The fore plane is longer than a jack, usually about 22 inches. Its

body is also wider, and its iron is ground square. The fore plane rides over the ripples created by the jack plane, trimming high spots with each successive pass. Eventually, the chip starts to discharge in long thin continuous shavings.

# Jointer

The largest wooden bench plane is the jointer. Its body is usually 26 inches long. The jointer made square edges for gluing boards together, and was used for all the purposes as the metal plane that bears the same name.

# Smooth Plane

The smoothing plane, or more simply the smooth plane, is a short plane with a wide sole that is used for general work or for smoothing small pieces. (See Illus. 49.)

Smooth planes are only about 8 inches long. Their mouths are wide, but their stocks are usually narrower at the toe and heel so that they will be easier to hold. Their shapes resulted in the colloquial name "coffin" plane.

Most smooth planes are used with the right hand placed on the heel and pushing from behind. The exception is the razee smooth plane that has a closed tote.

*Illus. 48. A group of 19th-century wooden bench planes. Clockwise from front right they are as follow: a smooth plane by J. Kellogg (Amherst, Massachusetts); an unmarked jack plane; a fore plane by Greenfield (Massachusetts) Tool Co.; and a jointer, also by J. Kellogg.*

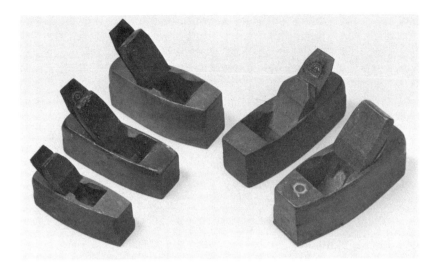

*Illus. 49. Although these factory-made wooden smooth planes all have a similar coffin shape, they are of different sizes. Clockwise from the left, they have the following maker's marks: unmarked except for "Made in England"; Scioto Works (Ohio Tool Co.); M. Wilson (New Castle, England); J. Kellogg (Amherst, Massachusetts); and Wm. Marples and Sons (Sheffield, England).*

# Selecting and Reconditioning a Plane

Bench planes were made in such large quantities that many good, sound ones are still available, and you will have no trouble finding the ones that you want. Still, most of the wooden bench planes you will see on the second-hand tool market have either been worn out by the woodworkers who owned them or have been ruined by abuse and neglect. Be selective.

Before purchasing a plane, examine the stock. It must be sound. Do not buy one that has large checks in the end grain, dry rot, or has been attacked by boring insects.

Many planes have a fracture in one of the cheeks, usually in a line that roughly corresponds to the angle of the bed. (See Illus. 50.) This results from too much force being exerted when the iron is adjusted laterally. Previous owners may have tried to repair the break with nails and screws, but this only makes the condition worse. Usually, a small crack is not by itself a problem, but a large crack cannot be easily repaired.

After you have selected a plane, remember that dust and grime collect under and around the iron and wedge and should be removed before you tune it or sharpen its iron. Some of this material is loose and you can remove it with your fingers or by blowing it off. More stubborn grime contains pitch and can often be dissolved with mineral spirits. Keeping the mouth free of this accumulation should be a regular part of your maintenance procedures for the plane.

A tool that is used to create a perfectly flat surface obviously must itself be perfectly flat. Whether a plane is made of metal or wood, the sole usually has to be flattened before it will work well.

*Illus. 50. This fracture in the cheek of a wooden razee smooth plane was probably caused by too much lateral adjustment.*

The sole of a wooden bench plane is often worn. It tends to wear out most at the right-hand corner of the toe, the left corner of the heel, and just in front of the mouth. The sole can also be twisted, a phenomenon called wracking. Wracking is usually the result of seasonal movement of wood rather than wear.

Test for wracking by using a pair of straight-edges called winding sticks. (See Illus. 51.) Set one on the heel and another across the toe. Sight down the plane. If the tops of the winding sticks are not parallel, the sole is wracked.

The solution for wear and wracking is to flatten the sole. I do this with a 24-inch jointer plane,

*Illus. 51. Check a bench plane's sole for wracking with a pair of winding sticks.*

*Illus. 52. Before jointing a wooden bench plane's sole, use a try square to determine if it is still at a right angle to the stock. Also check the sole during jointing and when finished. The plane shown here is out of square.*

*Illus. 53. Joint a plane's sole on a jointer held upside down in a vise. Make even passes; do not lift either the heel or the toe.*

using the same procedure that coopers used for centuries with a special type of jointer. First, adjust the iron laterally so that it cuts a shaving of the same thickness at both sides of the blade. Then adjust the cut so that it is paper-thin and place the plane in a vise upside down, with the toe facing towards you.

Next, loosen the plane's wedge and remove its iron. Lay a try square on the side of the plane and across its sole. This will tell you whether the sole is still at a right angle to the stock. (See Illus. 52.)

Pass the sole of the wooden plane over the sole of the upside-down jointer. Before each pass, place the toe just in front of the jointer's mouth. (See Illus. 53.) Using your left hand to hold the soles of the two planes tightly together, push with a steady, unbroken stroke. Be careful that the wooden body does not rock as you plane its sole.

By looking at the sole of the wooden plane, you can quickly identify the high spots. (See Illus. 54.) The jointer has shaved them, so they are lighter in color than the low spots. Repeat the process until the low spots have been removed and the entire sole is fresh wood.

Check to be sure that the sole of the wooden plane is square with the sides. If not, you will have to make more passes. Tip the stock slightly so that you are planing just the high side. Stop often, using the square to test your progress.

This is probably not the first time your plane's sole has been flattened. Because the mouth is the intersection of the two inclined surfaces (the wear and the bed), it becomes slightly wider each time the sole is resurfaced. (See Illus. 55.) Remember,

*Illus. 54. After you have passed the sole of a wooden plane over the sole of a jointer a couple of times, you will note that the high spots have been shaved and are lighter in color than the surrounding wood.*

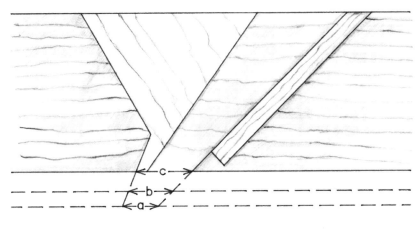

*Illus. 55. Each time the sole of a wooden bench plane is resurfaced, the mouth becomes slightly wider. a shows the original width, b represents the second width, and c the third.*

when a chip is being cut it is held down by the sole in front of the mouth and is forced up the throat, where it is broken by the chip breaker. As the mouth grows wider, the chip is allowed to lift in front of the cutting edge, resulting in more tearing. The solution to this problem is to close the mouth with a patch. First, cut the patch from a scrap of hardwood such as maple, ensuring that it is slightly longer than the mouth. Set the patch on the sole so that one edge is placed against the back edge of the mouth (the edge formed by the intersection of the sole and the bed). Now, trace around the patch with a sharp scribe. (See Illus. 56.)

*Illus. 56. Trace the outline of the patch with a sharp edge. I use a hobby knife. Like the plane, the patch shown here is made of beech. Note that the shape of the patch is similar to the stock's coffin shape.*

Use a sharp chisel to carefully cut an inlet that's the shape of the patch. (See Illus. 57.) The inlet does not have to be much more than three-eighths of an inch deep. Spread an even coating of glue on the inlet's bottom and edges. The type of glue you use does not matter.

If you have cut the inlet carefully, the patch should fit tightly. I use a vise to press the patch into the inlet. (See Illus. 58.) A large hand screw will work as well. In this case, you may want to place a shim over the top of the throat to avoid damage to the abutments.

Keep the patch clamped until the glue cures. Then chisel off the bulk of the waste and make a final pass over the jointer to ensure that the patch is flush with the sole.

Now, you have to reopen the mouth. Begin by working a saw blade between the patch and the bed. (See Illus. 59.) Enlarge the saw kerf with a chisel. (See Illus. 60.)

Put the iron into the plane to see how much more of the patch needs to be shaved away. Repeat this until the cutting edge projects from the mouth. The distance between the cutting edge and the patch should be no greater than the thickest shaving you will want to pass through it. This depends on the plane's use. A jack plane will have a wider mouth than a smooth plane.

When you are finished, take a moment to bevel the patch's inside rear edge to the same angle as the wear. The plane should now work as well as it did when new. (See Illus. 61 and 62.)

Let's look at the reconditioning methods I used on the most battered plane I could find. Normally, I would not have wasted my time on a tool that had this many problems. However, reconditioning it proved to be a helpful experience because I encountered and was able to deal with problems you might have to face someday; following is a review of how I solved these problems. Remember, any bench plane, whether a fore, trying, jointer, or smoothing plane, that experiences any of these flaws can generally be corrected in the same way.

The plane had the following problems: Its tote had been knocked off, and its wedge was broken. (See Illus. 63.) It had apparently been stored in a barn, as it was grey, covered with dust, and completely dried-out. Some check cracks had opened in the end grain, but the stock was still sound. Its sole was badly worn and the mouth was too wide. Its iron had been ground so much that very little steel remained.

The plane was made in the late 19th century by E. C. Rings, who owned a factory in Ringsville, Massachusetts, a small hamlet just west of Northampton. It had been owned by a woodworker named H. Cate, and he branded his stamp on both ends.

The first step in reconditioning Mr. Cate's jack plane was to make a replacement wedge. Although broken, the original was still with the plane, so I had no problem determining the proper shape. However, 19th-century designs were so

*Illus. 57. Excavate the inlet for the patch with a chisel. Cutting cross grain, as shown here, is called paring.*

*Illus. 58 (above left). I use a vise to press the patch into the inlet and to hold it in place while the glue dries. Illus. 59 (above right). Use the toe of a backsaw to open a mouth that has been closed by a patch.*

*Illus. 60. Carefully enlarge the mouth with a chisel.*

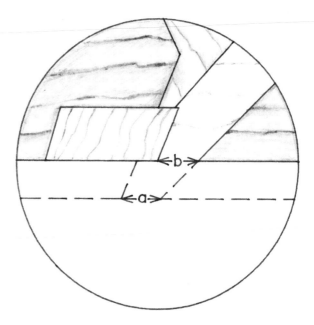

Illus. 61. When repeated resurfacing opens the mouth too wide, the solution is to close it with a patch. The dotted line represents the original sole. a represents the mouth's original width. The patch should create an opening of the same width, as shown at b. Be sure to bevel the back of the patch to the same angle as the wear.

Illus. 62 (above left). Next, flatten the sole and patch on a jointer plane. The new wood in this patch is barely distinguishable from the freshly planed surface. Illus 63 (above right). When I came across this jack plane, made by E. C. Ring of Worthington, Massachusetts, it had the following problems: Its tote had been snapped off; its wedge had a broken leg; and its iron was worn out from repeated grindings. The plane had been stored in a barn and its finish had turned a sickly grey. I have since reconditioned it.

standardized that you can copy from any similar plane.

The wedge has two legs that taper at the angle formed by the top of the iron and the abutments. When the iron is in the plane, these two legs have to fit snugly along their entire length. A high spot on the legs will act as a pivot point and allow the iron to rock in its bed. This will cause chatter. (See Illus. 64.) If the legs are at too steep an angle, the abutments will only grip them at the top. Not only will the blade chatter, the wedge will exert

pressure at that one spot and may break off the tops of the abutments. Consequently, I cannot stress too strongly the need for the wedge to fit properly.

I began by sawing out a blank that was the same width as the throat and roughly as long and as thick as the finished wedge. This makes it easier to form the taper correctly. Also, if you misjudge and have to discard the first attempt, you waste less work.

Next, I sawed out the space between the legs

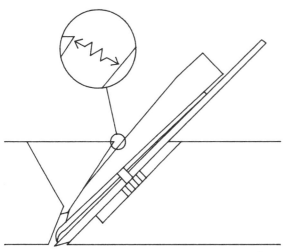

Illus. 64. *If a wedge's legs have a high spot or do not have the correct taper, the iron might rock in its bed, and can possibly chatter.*

Illus. 65. *This wedge is being made from a piece of beech. The old wedge served as a pattern. The blank will be planed until it fits snugly under the abutments, and then the legs will be sawn out.*

with a coping saw. (See Illus. 65.) I then formed a steeper incline between the legs to allow the chip to discharge more easily. (See Illus. 66.) Then I sawed the top of the wedge to its typical coffin shape and made a slight chamfer on all the visible corners.

I sanded the completed wedge with 220 grit paper to soften all of its surfaces. This simulates the wear an old wedge receives while it is being used. As the new wedge begins to darken, this artificial wear will help disguise the fact that it is a replacement. The purpose is not to deceive anyone. Wooden planes have an aesthetic appeal, and an obviously replaced part detracts from that appeal.

Illus. 66. *Finally, the incline between the legs is made with a chisel, and the inside edges of the legs are bevelled. These steps ensure that the wedge and throat eject the chip.*

The next step was to replace the tote, which I copied from a similar plane. I cut its outline with a coping saw, carved it to the finished shape with chisels (Illus. 67), and then sanded it smooth.

It is critical that the tote fit tightly into the stock, as it is only held with glue. No matter what else you might attempt (screws, nails, etc.), nothing will work as well as a good, tight joint.

Next, I jointed the sole and patched the mouth in the manner described on pages 51 and 52. In Illus. 68, you can see how much the sole had worn.

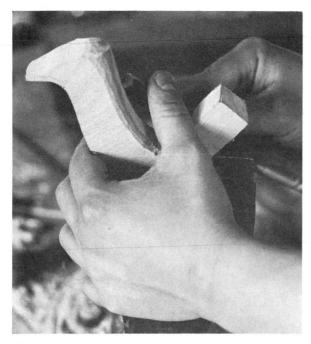

*Illus. 67. Most of the tote on the plane was missing, so this part was copied from another plane. Factory-made planes were very standardized, and it is probable that the copy is the same shape as the original tote.*

*Illus. 68. After closing the sole with a patch, I jointed the sole on an inverted jointer plane. This was not the first time the sole had been resurfaced. In fact, the stock had developed a distinct wedge shape; it was lower at the toe than at the heel.*

The stock was once a uniform thickness from toe to heel, but in Illus. 68 it has a distinct wedge shape, narrow at the front. This is understandable, since the user tends to lean on the front of the plane with his left hand. Friction can be reduced by waxing the sole with paraffin.

Coincidentally, I had a spare iron that was the same size as the plane's original. It was even made by the same factory, the Providence (Rhode Island) Tool Company. It was also decorated with the same cut-out corners. The replacement had sufficient steel remaining to make a convex fore plane iron. I ground and sharpened as described on pages 57–59.

I do not like to strip the finish off tools, but this one was an exception. Remember, it had been stored in a barn or cellar and was dirty-grey. I removed what little finish it had with a commercial paint remover that I brushed on and allowed to set. (See Illus. 69.) I scrubbed away the dissolved finish with fine-steel wool. This resulted in a brown patina that was able to form even on such a weathered plane. I was very careful to thoroughly clean from the throat any stripper that had flowed into it. Stripper that isn't removed will eventually harden and hold the wedge as tenaciously as glue.

Next, I gave the plane a new finish. I use five parts boiled linseed oil to one of turpentine. I brushed the oil on, allowed it to set for half an

*Illus. 69. The old grey finish was removed with paint stripper.*

hour, and wiped it dry. *Be sure to dispose of oil-soaked rags immediately, as they can combust spontaneously.*

This once-battered fore plane now looks as good as it ever will. One might even doubt that it was the same plane I purchased before I reconditioned it. Just to verify that my work was as successful from a practical standpoint as it was aesthetically, I ran the plane over a piece of rough-sawn pine. (See Illus. 70 and 71.) It performed as well as it ever did for H. Cate. I will retire the plane because, like most woodworkers, I am proud of my tools, and prefer to work with the best ones I can find.

# Sharpening the Iron

Make sure that the iron has sufficient steel left to be properly sharpened. The cutting edge should still be well away from the screw slot. If it is not,

*Illus. 70. The plane now works as well as it ever has.*

a blacksmith can weld a new piece of tool steel to a worn-out blade. However, bench plane irons are so common that this sort of repair is not worth the money and time it would require.

If you are grinding a straight cutting edge for a jointer, check the edge with a try square. (See Illus. 72.) If the edge is not square with the sides of the iron, make it so when you grind.

I spray the side opposite the bezel with layout fluid (see Chapter 8). Once the fluid is dry, use a try square and a sharp point (such as a scratch awl) to trace a line that is at a right angle with the sides of the cutter. (See Illus. 73.) Grind to this line.

If you are grinding a jack plane iron, follow the shape of the preexisting curve. If you are making a replacement iron, spray the side opposite the bezel with layout fluid and trace the curve you want.

The cutting edge on the smooth plane looks as if it is straight, but it actually has a very slight crest. This, too, can be traced in the layout fluid.

After grinding, check the side opposite the bezel. It must be perfectly flat and contain no pitting or other blemishes. Flatten this surface on the lapping table as explained in Chapter 7. Chapter 8 explains how to hone it to a razor edge.

When lapping a bench plane iron, flatten just the area that is steel up to an inch or two from the cutting edge. Keep the upper end of the iron away from the abrasive surface. (See Illus. 74.) Not only would it be a waste of effort to flatten the whole back of the blade, you could damage or even remove the maker's mark. This would lower the tool's value.

When sharpening the iron, remember that the edge of the chip breaker usually needs attention as well. If the edge where the chip breaker meets

*Illus. 71. The work done to the plane has turned it into a respectable tool. Through time and use, the new parts will slowly darken to match the old wood.*

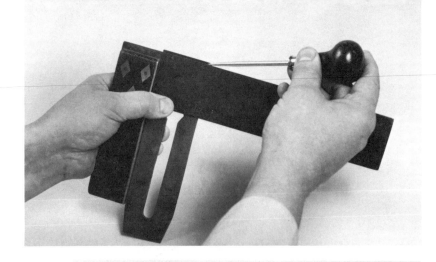

*Illus. 72. When sharpening a cutter for a jointer or smooth plane, make sure that the edge is square with the sides. If not, spray the side opposite the bezel with layout fluid and trace a line with a scratch awl. Grind to this line. When making a jack plane cutter, first trace the shape of the curved cutting edge in the layout fluid.*

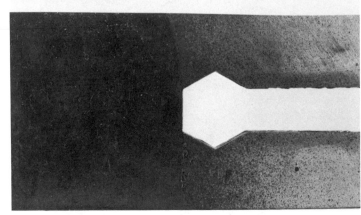

*Illus. 73. Grind the cutting edge to the line traced in the layout fluid. Here the line has been placed back farther from the edge than it needs to be so that it would be more visible.*

*Illus. 74. A clear maker's mark increases a tool's value, so when flattening a cutter hold it so that the mark is protected.*

the upper surface of the iron is chipped or rounded, the shaving can become caught underneath it and choke the plane. To ensure that it fits tightly, without any gap, straighten this edge on the lapping table.

The chip breaker's edge should be set less than ⅛ of an inch behind the iron's cutting edge. When its setscrew is tightened, the chip breaker gives the plane blade a slight cup that helps stiffen it.

# Using a Plane

Woodworkers who use metal planes argue that the cam-operated lever cap—the horizontal and lateral adjustments of the iron—and the adjustable frog are major advantages metal planes have over those made of wood. Craftsmen who use wooden tools dispute this. A wooden plane can be set more quickly and just as accurately. I prefer to use a cobbler's hammer to do this because its metal head provides the sure blow that is required to place and adjust the iron, while its wide face will not damage the wooden parts.

Loosen the iron in a wooden plane by striking the heel. (See Illus. 75.) Make sure that the blow from the hammer or mallet is square on the heel so that it doesn't damage the plane.

The same results can be obtained by striking the stock just in front of the throat. Some planes

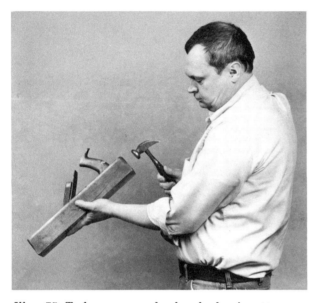

Illus. 75. To loosen a wooden bench plane's cutter, strike the heel. Some woodworkers strike the top of the stock in front of the mouth. If the plane has a strike button, hit that.

are equipped with a strike button in either of these two places. (See Illus. 76.) It is usually a hardwood plug with vertical grain, though it can also be iron, and even ivory or bone. The button's purpose is to absorb the blow while still allowing it to loosen the wedge.

When you wish to resecure the iron, set it into the throat. Place your fingers over the mouth and let the cutting edge of the iron rest on your fingertips. Insert the wedge so that it lies on the chip breaker and its legs slide under the abutments. Push the wedge with your hand. This will create enough friction to hold both it and the iron in place.

The iron can now be adjusted. Keep your fingers over the mouth so that you can feel the edge as the iron moves. If you lightly tap the back of the iron, you can feel the edge move forward. (See Illus. 77.) It should be advanced until it just barely projects out the mouth. Until you develop the ability to do this completely by feel, you can watch the cutter move by holding the plane upside down and sighting down the sole, toe to heel.

If you should misjudge and set the iron too deeply, retract it by tapping on the heel of the plane. The harder the tap, the more the iron will move backwards. This forward and backwards movement is called the longitudinal adjustment.

When you think you have achieved the longitudinal setting, make the lateral (side-to-side) adjustment. You can do this by feeling both sides of the cutting edge or by sighting down the sole. The iron is usually higher on one side than the other, and you can correct this by tapping on the

Illus. 76. Planes with strike buttons. The strike button could be made of a variety of materials such as end-grain hardwood, iron, ivory, or bone. A hammer or mallet was used to loosen the cutter. This protected the stock. Note the damage to the stock of the plane (right) that does not have a strike button.

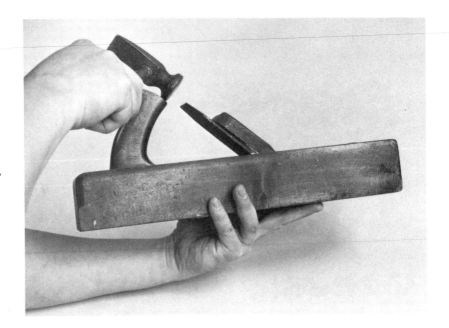

*Illus. 77. To advance a wooden bench plane's cutter longitudinally, tap the back of the cutter.*

side whose set is too heavy. (See Illus. 78.) If you overcompensate, tap the iron back from the other side.

When you have both the proper longitudinal and lateral settings, tap the back of the wedge for the final tightening. Do not strike the wedge too hard. It only has to keep the iron in place while the plane is in use. Too much pressure risks crushing the end of the wedge or the abutments or cracking the cheeks. A wedge that is too tight requires more force to knock it loose. This force can damage the plane's body.

Now, test the plane on a piece of wood. If the plane does not cut, or the chip is too fine, advance the iron by tapping the upper projecting end of the iron. If the chip is too thick, retract the iron by tapping the rear of the stock or the strike point. Test again. When the plane works the way you want, strike the wedge one last time to make sure it is locked.

Someone who has never adjusted a plane iron the way I've just described may think it more time-consuming and less exact than adjusting a metal plane. However, the ability to adjust with accuracy is developed in a very short while and, eventually, will be much faster than finding and turning knobs and levers.

I store my planes on a shelf under my bench. I have tacked a thin strip of wood along the back of the shelf and set the toe of the plane on this strip. This raises the cutting edge off the shelf.

*Illus. 78. To laterally adjust a wooden bench plane, lightly tap the side of the cutter. To move it in the opposite direction, tap the other side.*

# 10
# Metal Bench Planes

The metal bench plane as we now know it was first developed by Leonard Bailey of Winchester, Massachusetts, just after the American Civil War. Bailey took advantage of technological advances that made it possible to cast iron more precisely. Ironically, he was looking for a way to make planes less expensively, rather than to make better planes.

In the 1870's, Bailey sold his patent rights to a company called Stanley Rule and Level, which continued to make his line of planes under the Bailey name. In 1870, Stanley Rule and Level was making 11 different iron planes using Bailey's ideas.

Over the following decades, Stanley made only minor improvements to the metal bench plane. During that time, many other companies in both England and America made planes based on Bailey's patterns. In fact, if old Leonard could flip through a modern tool catalogue, he would easily recognize the modern versions of the plane he invented more than a century ago.

## Plane Parts

Metal planes have two major parts: the body (also called the bottom) and the frog. Each part is discussed below.

## Body

The lower surface of the cast-iron body is still called the sole, and the opening in the sole, through which the cutting edge projects, is the mouth. On the body's upper surface, just in front of the mouth, is a projecting ridge which acts as a rudimentary wear. The wear is not as important on a metal plane as it is on a wooden one. The

distance the shaving travels while being ejected is shorter, and there is less risk of choking.

The body has raised sides that curve up to their highest points adjacent to the iron. These points are called the cheeks. The body has a raised boss in the toe, and another in the heel. These bosses are drilled and tapped and are used to secure the knob (front) and the tote (rear) to the plane's body. The knob and tote are the only wooden parts on the plane. The other parts are made of cast iron, steel, or brass.

## Frog

The frog is a three-sided piece of machined cast iron that is fastened to the body, just behind the mouth. When you remove the lever cap and iron you reveal the two screws that hold the frog to the body. If you loosen these two screws, you can remove the frog and expose the raised platform on which it sits. The platform usually has two machined surfaces that help hold the frog perfectly rigid. On planes made with its "Bedrock" trademark, Stanley did use a slightly different system that involved a single wide surface. This made the frog even more rigid. For this reason, Bedrock planes are desirable and worth buying if you find one that is in good condition.

The frog holds the cutter assembly and allows it to be adjusted in three different directions. To understand how the frog functions, it is necessary to consider the three directions in which an iron is set: longitudinally (advance and retract), laterally (side to side), and parallel (back and forth movement of the frog). The third direction is used only on metal planes, and was not possible on wooden planes.

The frog's front surface is called the bed. It is a 45-degree incline to which the cutter is secured. A

## A Bench Plane

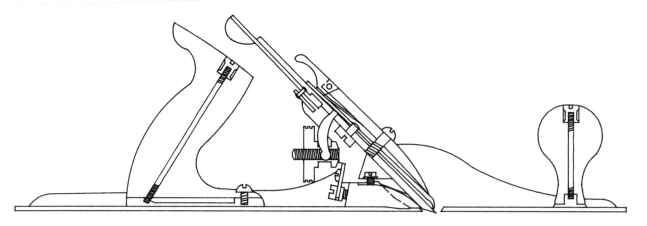

*Illus. 79. A cutaway view of a #5 jack plane.*

## A Bench Plane

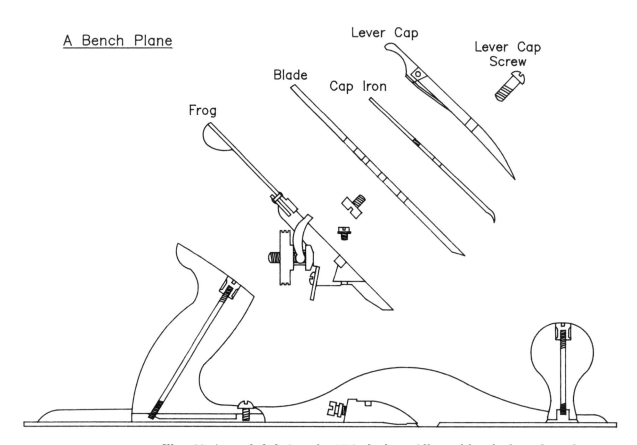

Lever Cap

Lever Cap
Screw

Blade

Cap Iron

Frog

*Illus. 80. An exploded view of a #5 jack plane. All metal bench planes have the same parts. Earlier planes (made before 1914) may not have a frog adjustment screw, and very early Bailey planes do not have a lateral adjustment lever. However, these planes are not common, and you should leave them for collectors because a lateral adjustment is a very important feature.*

roundheaded machine screw, called the cap screw, projects from the bed. The cutter/chip breaker and lever cap have holes in them, and all three fit over this screw.

The cam lever cap was devised by Bailey (patented in 1858) to hold the iron and chip breaker firmly against the bed. (See Illus 81.) The cam lever is in the cap's upper end, and when pushed down it locks the cap in place. The lever cap eliminated the need for a wedge. It also has other advantages that will be described later.

You can release the cap and the iron/chip breaker assembly very quickly by lifting the lever. This loosens the iron so that it can be removed for cleaning or sharpening. The cam lever eliminates the need for a mallet or hammer

*Illus. 81. The cam lever cap (shown here with the lever up), invented by Leonard Bailey, eliminated the need for a wedge to hold the cutter.*

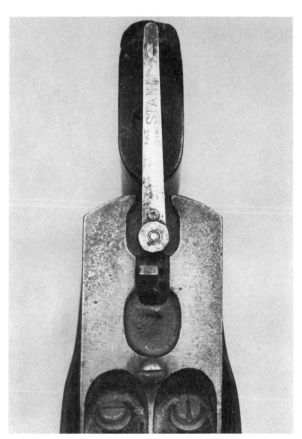

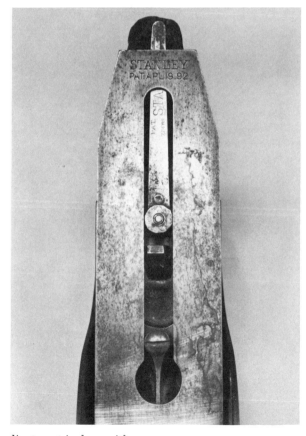

*Illus. 82 (above left). On a cast-iron bench plane, lateral adjustment is done with a lever. The lever is secured to the frog with a rivet; the lever pivots on the rivet. The two screws that secure the frog to the cast-iron body are also visible here. Illus. 83 (above right). The cutter is moved laterally by a friction-reducing wheel (newer planes have a lip) on the end of the lever, which fits into the cap screw slot.*

when you are adjusting the plane. Instead, you need a screwdriver.

To laterally adjust the iron, use the lever fitted into the bed's upper end, just under the iron. This lever is secured with a rivet, on which it pivots. (See Illus. 82.) On the lower end of the lever, about a half inch below the rivet, is a wheel (on earlier planes) or an upturned lip (on later planes). (See Illus. 83.) The wheel (or lip) fits into the slot in the iron. As you pivot the lever, the wheel (or lip) pushes the iron side to side. This raises or lowers the corners of the cutting edge.

To laterally adjust the cutter, move the lever in the opposite direction from the motion of the blade. For example, if you push the lever to the right you will lower the cutter's left corner.

To longitudinally adjust the iron, turn a brass nut in back of the frog. (See Illus. 84.) The nut is mounted on a left-hand threaded shaft. It has a neck or groove that fits into the open yoke of a Y-shaped lever. The Y-shaped lever is fastened so that it too can pivot. Its upper end fits into a notch in the chip breaker. (See Illus. 85.) As the nut is turned, it moves up or down the threaded shaft, pivoting the Y-shaped lever. In turn, the iron is either advanced or retracted, increasing or decreasing the thickness of its cut.

The third adjustment is parallel to the sole: the

*Illus. 84. To longitudinally adjust the cutter, turn the knurled brass knob behind the frog. The knob is connected to the Y lever, which is connected to the chip breaker. To advance the cutter, turn the knob clockwise; to retract it, turn the knob counterclockwise. Note the frog adjustment screw below the brass knob. Stanley began to add this feature to its planes about 1914. It eventually became a standard feature on all iron bench planes.*

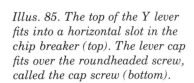

*Illus. 85. The top of the Y lever fits into a horizontal slot in the chip breaker (top). The lever cap fits over the roundheaded screw, called the cap screw (bottom).*

forward and backwards movement of the frog itself. This opens or closes the mouth. The frog is held in place by two screws under the cutter assembly. If these screws are loosened, the frog can be moved forward or backwards. This places the cutting edge closer to, or further away from, the mouth's front edge and makes the mouth wider or narrower.

On planes made before 1910, you adjust the frog with your fingers. On later planes, an adjustment screw was added to the lower rear of the frog. In this case, the two setscrews still have to be loosened, but you can move the frog by turning this adjustment screw with a screwdriver.

The cutter on a metal bench plane is not tapered (see Chapter 9), but has a uniform thickness and is called a parallel iron. This type of cutter was less expensive than the older tapered irons, since the cutter could be stamped rather than forged. These irons are made of tool steel rather than wrought iron with a forge-welded layer of steel. This thin type of iron is less able to resist chatter than a thicker tapered iron, but this problem is eliminated by the lever cap. The cap's straight front edge creates even pressure across the entire cutting edge (rather than just the sides as does a wedge).

The soles of some metal bench planes are corrugated with at least ten grooves that run from the toe to the heel. (See Illus. 86.) Explanations for the grooves vary, depending on the source. One explanation is that they were used to reduce friction, another that they prevented suction; both friction and suction make the plane more

*Illus. 86. Some woodworkers claim that a corrugated sole reduces friction, making the plane easier to push.*

*Illus. 87. It is easy to identify the various sizes of these Stanley iron bench planes, as the number is usually cast into their toes. To date the plane, however, you might need an awareness of the changes the company has made to its planes over the years.*

*The plane on the far right is a #3 smooth plane that has an 1892 patent date stamped into the cutter. The plane had to be made after that date. In 1905, Stanley redesigned the frog. Since this plane has the old-style frog, it was made before 1905.*

*The Bailey #4 smooth plane, second from right, and the #7, far left, both have a new-style frog, and were thus made after 1905. Both do not have a frog adjustment screw, which indicates they were made before 1914.*

*The #5, third from right, is a later model plane, but it still has the name Bailey cast into its toe. It has a taller knob that was introduced about 1922. The word Stanley shown cast into the lever cap was not used until after 1925, and the kidney-shaped hole in the lever cap was patented in 1933. The raised ring under the knob was added about 1936, so the plane had to have been made after that date.*

difficult to use. In my experience, the only difference the grooves make is that it takes less time on the lapping table to flatten a corrugated sole.

Metal planes come in roughly the same lengths as their wooden counterparts and, like the wooden planes, are classified as smooth, jack, fore (try), and jointer planes. (See Illus. 87.) Metal planes made by Stanley were classified by a system of numbers that identified the different sizes of planes. For example, a number 1 plane is a small smooth plane only 5½ inches long. The planes numbered from 2–4 are also called smooth planes and range up to 10 inches in length. The number 5 plane is called a jack plane and is 14 inches long. The number 6 plane is a fore plane, and is 18 inches long. The number 7 and 8 planes are both jointers, and are, respectively, 22 and 24 inches long.

Over the years, Stanley has published numerous catalogues of the tools that it has manufactured. Tool collectors and dealers have reprinted these catalogues so that a detailed chronology of these planes, and the dates of the changes and improvements made to them, is available. (See the Bibliography on page 254.)

For several decades after Stanley purchased Bailey's patent rights, its planes bore the name of Bailey. For this reason, these older planes are commonly called "Bailey Patent" planes.

Other companies made similar-sized planes, and these too are commonly found on the second-hand market. The company's name is usually cast into the cap or the body.

Because metal planes are made of cast iron, they are susceptible to rust. To help protect the tool, the inside of the plane's body and the exposed surfaces of the frog were covered with a shiny finish of linseed oil and asphalt pitch, called japanning. On early Stanley planes, this japanning was usually black.

The outside of the body, the sole, the cap, the chip breaker, and the iron were machined smooth, but left unfinished (on later planes, the cap is often plated). The tote and front knob were lacquered.

Second-hand planes, like antique tools, are increasing in value. To protect your investment, remember that an original finish is desirable and increases a tool's value, even if that finish is not perfect.

# Selecting and Reconditioning a Plane

Over the last 100 years, so many metal planes have been made by Stanley and other companies that these tools are very common on the second-hand market. You should have no trouble finding them no matter where you live. Therefore, only buy tools that are still in very good condition. Don't buy one that has any parts that are missing or broken.

Before purchasing a second-hand bench plane, flip up the cap's cam lever. This will allow you to remove the cap and the cutter assembly so that you can examine the working parts. Separate the cutter and the chip breaker. Of course, to do all this you will need a screwdriver. A travelling kit is very valuable in a situation like this.

Examine all these parts for damage and rust. A light layer of rust can be easily removed, but heavy rust will cause pitting. Check the cap. The cam should operate smoothly, and the spring that holds the cam in place should still be stiff enough to do its job.

Look carefully at the frog and, if possible, take it out of the plane. Make sure that the screws that hold it in place are not worn or stripped. Look at the bottom of the frog and examine the machined surfaces on which it sets. They should be clean and undamaged. The same criteria apply to the mating surfaces in the plane's body. If these surfaces are damaged or badly rusted, the frog may not seat correctly and may rock when the tool is used.

Test the action of the longitudinal adjustment knob. The left-hand thread should not be too tight. If it is, determine whether it is just caked with grime or if the threads are stripped or damaged. On the other hand, if the knob is too loose, the threads may be badly worn and may not hold a setting.

Check the action of the Y-shaped lever. It should move freely, and the lever itself should not be damaged. If it is worn, it may have a lot of slack. This means that each time you make an adjustment, it will be necessary to turn the knob several times before it will engage the yoke.

The lateral adjustment lever should move freely, but without excessive play. If there is a wheel on the lower end, it, too, should turn easily. The round-headed machine bolt (cap screw) over which the cap is fitted should be the original one, and its threads (as well as those of the hole in the frog) should be in good condition. Sight down the sole to see if it is excessively worn. The mouth should be clean, with no chips in the edges. The tote and front knob should be tight and unbroken.

The condition of the japanning will not affect how well the tool works. Nor would I be concerned about a small chip in the edge of the body; however, I would reject a plane whose casting was cracked. There are, of course, always exceptions; in these cases, apply the criteria outlined in Chapter 4 (page 18).

In the event that you have inherited a tool that has great sentimental value, you may want to put more time or effort into making it work than you would an average plane. In this case, there is an advantage and a drawback. The drawback is that remaking or repairing metal parts is often outside the experience and ability of a woodworker. Even when you have these skills, making or repairing parts can take more time than the job is worth. The advantage is that the parts were mass-produced and are usually interchangeable. You can cannibalize them from another plane as long as it is the same model, made by the same maker.

During the last several decades very cheap handyman versions of metal bench planes have been produced and sold by hardware dealers and through general merchandise catalogues. These tools are found on the second-hand market. To avoid buying these tools, just buy planes made by companies you know and recognize, such as Stanley, Sargent, Record, etc.

Once you have acquired a metal bench plane (new or old), you have to get it into working order. Start by disassembling the plane in the same manner as described on pages 63 and 64. Clean each part with mineral spirits to remove dirt and grime. Do not use harsher solvents, as they can damage the finish. You may have to use a toothbrush to reach into corners. Pay attention to the contact surfaces between the body and frog. These, too, must be cleaned.

There is often a buildup of pitch and resin on the front edge of the cap and on the cutter. This buildup will usually dissolve in mineral spirits as well.

Steel wool will remove light rust from unfinished surfaces. Try the 00 grade of steel wool. If it is too fine, use something more coarse.

Although the sole of a metal bench plane is much more durable than the sole of a wooden plane, it, too, will eventually wear until it is no longer flat. Of course, the soles of new planes are not flat either. So, put the frog back on the body and take the plane to the lapping table. (The frog creates stresses, which can affect the sole's flatness, so it has to be in place when the sole is flattened.) Adhere at least two sheets of emery cloth or garnet paper (220 grit) to the table. Use three sheets if you are lapping the sole of a jointer plane. Butt the sheets tightly together so that you have a surface that is 20–30 inches long.

Hold the plane so that its sole is on the paper. (See Illus. 88.) Make several steady strokes; make sure that the plane stays flat on the table and does not rock. The high spots will soon be abraded enough to show new metal. (See Illus. 89.) These spots are a lighter color than the darker low areas. Typically, the sole will be worn most just in front of the mouth. It may also be worn on the front right corner and the rear left corner.

The sole should be lapped until it is a single color. Make sure that each stroke begins with the entire sole in contact with the paper, and do not lift the plane until the stroke is finished.

In hand woodworking, many small pieces of wood are jointed on a device called a shooting board. (See Illus. 90.) (Shooting is the process of squaring a board's edge with a plane.) To "shoot" these edges, lay the plane on its side, rather than on its sole. This means that the plane's right cheek has to be square with the sole. If you are going to use your plane on a shooting board, check for square with a try square. If the cheek and sole do not form a perfect right angle, return to the lapping table. As you make each stroke, bias the pressure towards the high edge. This way you can use the abrasive surface to slowly bring the cheek into square. Check regularly until the job is finished.

Before reassembling the plane, lightly oil the moving parts and work them until they move easily. As each part is cleaned, spray it lightly with a rust inhibitor, such as WD-40®.

*Illus. 88. Lap the sole of an iron bench plane on a lapping table. Keep an even pressure on the sole at all times. Here, I am using garnet paper. Note that it has been butted together to produce a long narrow surface.*

*Illus. 89. The sole of this plane has been partially jointed on a lapping table. The high spots, which have been abraded by the sandpaper, show as shiny metal. Obviously, the sole is far from being flat and would never work well if left as is. Soles of new iron planes usually need jointing, too. Flattening should also be part of a plane's maintenance. I rejoint my iron planes about once a year.*

*Illus. 90. Using a #5 plane on a "joint and square" shooting board.*

# Sharpening the Iron

The next step is to sharpen the iron. Follow the procedures explained in Chapter 8. When grinding the edge, remember that the iron is much thinner than the iron on a wooden plane, and is, therefore, easier to burn. When this happens, you will see the telltale blue that indicates that the spot has overheated and lost its temper. Soft spots will not hold an edge. Because these cutters are so thin, I cool them regularly in water when grinding.

Before honing the edge, be sure to flatten the side opposite the bezel. Also hone the front edge of the chip breaker before assembling it with the iron. As on a wooden plane, it is important that this edge fit tightly on the iron so that it can prevent chips from jamming under it as they travel up and out of the plane.

Cutters are not always in good condition, and sometimes have to be replaced. Replacement irons are readily available through many tool catalogues and will fit most metal planes you will find on the second-hand market.

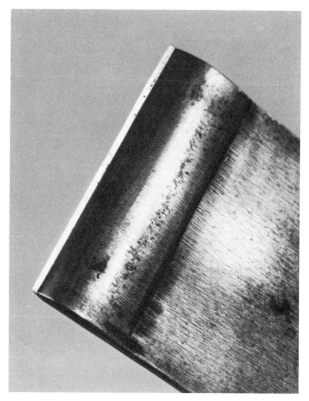

*Illus. 91. The chip breaker should be about 1/16 inch back from the cutting edge.*

# Using a Plane

Before returning the iron, take a minute to set the frog. This adjustment depends on how you intend to use the plane. If it is going to do fairly heavy work (removing thick shavings), set the frog back so that the mouth is fully open. For very fine work on figured woods, set the frog as far forward as possible without pressing the cutting edge against the front edge of the mouth. This will allow the tool to cut a tissue-thin shaving with an absolute minimum of tear out.

Once the frog is adjusted, tighten the screws that hold it in place. Do not overtighten them, as you risk warping the sole or damaging the casting.

Next, assemble the cutter and the chip breaker. The edge of the chip breaker should be about 1/16 of an inch behind the cutting edge. (See Illus. 91.) Place the cutter assembly in the plane, making sure that the tab on the upper end of the Y-shaped lever engages the notch in the chip breaker.

Place the cap over the cap screw so that it is resting on top of the cutter/chip breaker assembly. The cam lever must be turned upward. When you push the lever down, it should lock firmly. If it drops too easily, it will not hold the iron securely. If it is too tight, you risk breaking the lever or damaging the frog. Advance or retract the cap screw until the lever has the right amount of pressure.

Next, sight down the sole of the plane by holding the tool upside down with the toe forward. (See Illus. 92.) Use your free hand to work the longitudinal knob. As you turn the knob, you will see the cutting edge emerging from the mouth. You can also see whether or not the edge is level. If one side is higher than the other, push the lateral adjustment lever in the other direction until the edge is parallel with the sole.

Once you have become accustomed to the tool, you can adjust by feel rather than by sight. Once again, make sure that the edge is retracted below the sole. Hold the tool in one hand, with the fingers of that hand playing over the mouth. Turn the brass knob with your free hand. As the iron advances, your fingertips will feel the edge move. With repeated experience, you will be able to judge the set with just your fingers. You can also feel when the tool needs lateral adjustment.

Once you have the iron set where you think it should be, try the plane on the surface of a board. If the tool cuts too heavy a shaving, retract the iron. Advance it if the shaving is too thin.

If the tool is digging on one side, the iron does not have the proper lateral adjustment.

Do not use the same plane for both heavy and fine work. Buy a separate plane for each function. I often use a wooden jack plane for heavy planing and a Bailey Patent #4 smooth plane for finish work.

Once the plane is working and in regular use, you should periodically clean it to minimize the buildup of grime.

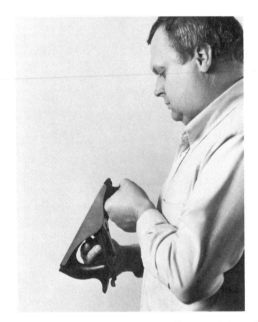 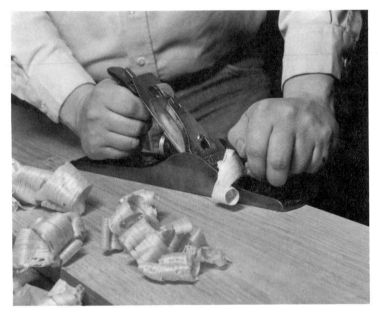

*Illus. 92 (above left). When adjusting a metal bench plane, sight down the sole. With one hand, turn the longitudinal adjustment knob or move the lateral adjustment lever. You will be able to see the cutting edge move. Illus. 93 (above right). Using a razor-sharp, well-tuned bench plane is one of the most pleasurable aspects of woodworking.*

# 11
# Transitional Planes

After the American Civil War, plane-making factories started to produce the cast-iron planes that were discussed in Chapter 10. Most of the same factories also made what are today called transitional planes. These tools are a hybrid between the older wooden bench plane and the cast-iron plane. They have a wooden stock, and most of the other parts are made of metal.

The inventive Leonard Bailey was also instrumental in the development of this type of plane. Along with his other patents, Bailey also sold rights to his wooden-bottomed planes to Stanley Rule and Level Co., and by 1870 Stanley was making 17 different types of wooden-bottomed planes. As other factories began to produce similar planes, the designs of transitional planes became relatively standard. As a result, no matter who the manufacturer is, all these planes look very much alike. (See Illus. 94.)

The word transitional is misleading, because these planes were not really a step in the evolution of the wooden plane to the modern metal plane. Bailey developed both forms at about the same time. Transitional planes were perhaps a less-expensive product that still offered the features woodworkers liked in the new metal planes (for example, the adjustable frog and the quick cam-lever cap). They also may have been a way to overcome conservative attitudes among woodworkers who retained a preference for wooden planes and looked with suspicion on the new, all-metal types.

## Plane Parts

The most common transitional plane you will see on the second-hand market has a beech stock that's only about $1\frac{1}{2}$ inches thick. Like a wooden bench plane, this shallow wooden body has a throat and mouth cut into it. (See Illus. 95.) There are also cheeks and a short wear. There are no abutments because the iron is not secured with a wedge, but with a lever cap. The back of the throat forms a short bed to support the lower half of the cutter.

Transitional planes also have a frog, which supports the upper half of the iron. The frog is very much like those on metal planes. It, too, has a brass knob that is turned for longitudinal adjustment and (on all but the very earliest transitional planes) a lateral adjustment lever.

The frog is attached to the plane by two round-headed wood screws that can be loosened for adjustment. However, part of the bed is cut into the wooden stock, and is in a fixed position. As a result, the frog adjustment is not as effective as that on a metal plane.

Transitional planes also have a double iron which, of course, means one with a chip breaker. Depending on the maker, some transitional planes came with tapered irons, while others used the flat, modern iron. No matter whether the iron is tapered or flat, it and the chip breaker are secured to the frog by a cam release cap just like those found on metal bench planes.

The top of the stock is fitted with a cast-iron superstructure, called the top casting. This part looks very much like the leather upper on a shoe. The top casting forms a rim around the mouth and extends forward under the knob and back under the tote. It is screwed to the wooden stock. The wooden knob and tote are secured to the casting with screws (the knob screw passes through into the stock).

Transitional planes come in the same sizes as wooden bench planes, which are as follow: smooth,

*Illus. 94. The smooth plane, at far right, was made by Ohio Tool Co. (their #024), while the jack plane, in the middle, was made by Stanley Rule & Level Co. (their #27). The fore plane, at far left, has no mark; neither does its cutter. It was probably made by a known company, but was perhaps left unmarked so that it could be sold by a tool dealer under a private label. Note that it has been spattered with paint, though underneath the paint it retains its original finish.*

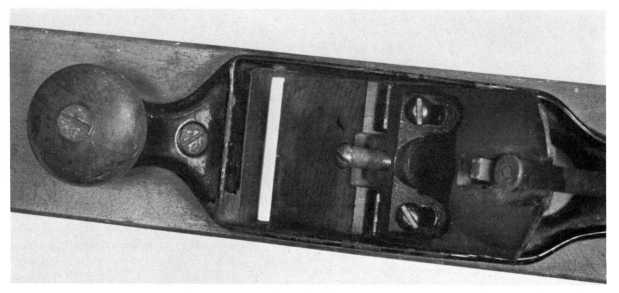

*Illus. 95. Transitional planes have an adjustable frog and mechanisms that allow for lateral and longitudinal adjustment. However, on an iron plane the bed is the frog's front surface. On this type of transitional plane, the frog only provides about half the bed. The rest is cut into the wooden body.*

jack, fore (remember, the term fore plane replaced try plane during the 19th century), and jointer. Like iron planes, transitional planes were also numbered by manufacturers. These numbers were usually stamped, along with the maker's mark, into the toe of the wooden stock. For instance, the Ohio Tool Company smooth plane shown in Illus. 94 is known as number 024. The Stanley Rule and Level Company jack plane is number 27. Note that the smooth plane does not

*Illus. 96. This type of smooth plane is also considered a "transitional" plane. Although it has a coffin-shaped wooden body, it also has a metal plate in front of the mouth. By loosening a screw on the top of the stock, you can slide the plate back and forth to adjust the width of the mouth. A similar device was used on block planes. (See Chapter 14.)*

have a tote. Many manufacturers also produced what was called a jack-handled smooth plane that looked very much like the wooden razee smooth plane.

The hybrid planes described above are what most people think of as "transitional planes." However, the term also applies to a number of lesser-known tools. Since this book only includes

tools with which I am familiar, most of these planes will not be discussed.

However, one of my smooth planes is another transitional type. Its wooden stock is coffin-shaped, and the iron is held in with a traditional wedge. A movable frog, as found on iron and transitional planes, allows you to adjust the size of the mouth, opening it for heavy work or closing it to cut extremely fine shavings.

This adjustment is not normally possible on a wooden plane because of the fixed bed. However, the English company (Wm. Marple & Sons) that made this plane included a feature that gives the wooden smooth plane an adjustable mouth. The plane has a movable metal sole in front of the cutting edge. A similar device is included on some cast-iron block planes (see Chapter 13). The movable sole is held in place by a bolt that runs through the toe. Its slotted head is visible in the top of the stock, in front of the throat. To open or close the mouth, use a screwdriver to loosen the bolt; this allows the metal sole to slide forward or backward. To retain the new position, tighten the bolt.

Transitional planes are the orphans of the second-hand tool market. They are of little interest to collectors, and most woodworkers are not familiar with them. As a result of this low demand, they are very inexpensive. I recommend buying, tuning, and using a transitional plane as a low-risk way to introduce yourself to second-hand tools. You will find that it works very well.

# Selecting and Reconditioning a Plane

Most transitional planes are inexpensive, so do not buy one that's in less than excellent condition. Since they are also common, you will not have to wait long to find one in such shape. Before purchasing the tool, examine it as you would both a wooden and a metal plane.

Examine the stock to make sure that it is sound. It should not be excessively worn or battered. Look at the sole to determine that the mouth is still tight. Many transitional planes have damaged totes with tops that have been broken off. Wait for a plane that does not have any of these problems.

Do not buy a transitional plane that is missing

any parts. Although these planes were mass-produced, you would have to salvage an exact duplicate to get the parts needed. The tools of different makers do not necessarily have interchangeable parts; nor do the tools even made by the same company.

Make sure that the plane has all of its original parts. The cap, chip breaker, and iron should all fit correctly, and the screws that hold the metal parts to the wooden stock should all be tight. Do not buy a plane if these screws are stripped. Make sure that the knob and tote are tight.

The wooden parts—the stock, the knob, and the tote—were all made of beech and were varnished at the factory. The metal parts were japanned, and are usually red or black. Most of the plane's finish should be original and intact. However, the plane is a woodworking tool, and it is unreasonable to expect the finish to be perfect.

No matter how good the condition of the tool, if you want to use it for fine woodworking you will almost always have to joint the sole. Do this in the same manner as you would the sole on a wooden bench plane.

Although you should not buy a transitional plane that is less than perfect, you may inherit a damaged plane to which you have a sentimental attachment. In spite of its low value, you may want to restore it. You can make repairs to the wooden parts because you are a woodworker. Follow the processes in Chapter 9. If you need to make either a new tote or knob, copy either the original or the same part from another plane of the same type. The knobs and totes of iron bench planes are made of rosewood rather than beech, but provide the same patterns.

You will find it more difficult to make repairs to the metal parts unless you also know how to work metal. If a tool with sentimental value has broken metal parts, you will probably have to salvage them.

# Sharpening the Iron

Before sharpening the iron on your transitional plane, think about the job you want it to do. If you want to use it as a jack plane for surfacing rough-sawn lumber, grind the edge to a radius. On a jointer, the cutting edge is usually straight, and on a smooth plane it is slightly crested. Transitional planes use the same cutters as wooden and metal bench planes, so when sharpening follow the advice in Chapters 6–8. The replacement blades sold through new tool catalogues will fit many transitional planes.

# Using a Plane

Secure the chip breaker to the iron the same way you would secure a chip breaker on a metal plane and return the assembly to the plane. Place the cap over the screw that holds it in place. As on a metal plane, the cam should not be loose, nor should it be so tight that it is hard to operate.

Feel the mouth while you advance the iron. You will discover that the longitudinal adjustment knob is very close to the stock and there is less room for your fingers than on a metal plane. Once the cutting edge has begun to protrude, you can set the lateral adjustment by sighting down the sole to see if either corner of the cutting edge is higher than the other, and then moving the lever behind the iron to adjust it.

# Specialized Planes

# 12
# Wooden Moulding Planes

Woodworkers have used mouldings to embellish furniture and architecture for hundreds of years. These decorative, three-dimensional bands, many with unfamiliar Latin names, have been used to add a special effect to homes and furniture. In better houses, different moulding profiles were used to make door and window architraves, cornices, chair rails and mopboards.

Today, all mouldings are made by machine. However, from the time of the Romans until the mid-19th century, mouldings were made with wooden moulding planes. English woodworkers were using moulding planes when America was first settled. These tools continued to be used by many woodworking tradesmen such as joiners, house wrights, cabinetmakers, and ship builders in both England and America for another three centuries. Even after World War I, wooden moulding planes were still being sold by large companies like Sargent and the Sandusky Tool Co.

For small shop woodworkers, there is no more efficient way to make mouldings than with a wooden moulding plane. In fact, the methods used today are more difficult and more complicated, and the results are worse.

Moulding planes were made in prodigious numbers, both in England and in America, and, therefore, can still be found in large quantities on today's second-hand tool market. As you look for second-hand tools to use for woodworking, you will come across many wooden moulding planes. They are available everywhere: in flea markets, yard sales, antique shops, and, above all, from specialized tool dealers.

If you acquire just a few basic moulding planes and learn to use them, you will discover how easy it is to make mouldings with them, and will probably begin to include more mouldings in your woodworking than ever before. You do not have to make reproductions of furniture to use moulding planes; mouldings look good on even contemporary furniture.

Many moulding planes look very much alike, but there is also a tremendous variety. Some planes make mouldings only a fraction of an inch wide, while others cut four- and five-inch cornices. Some planes produce a wide flat shape, while others have to be worked down into the wood an inch or more to form a narrow but very deep moulding. (See Illus. 97.) Some planes are so

*Illus. 97. Moulding planes come in a great variety of sizes, ranging from those that are very wide and shallow to those that are narrow and deep.*

small that they can fit in the palm of a hand, while others are so big that one worker alone does not have the strength to use them. In the latter case, a rope is attached to the front of the plane, which is pulled by one worker and pushed (and guided) by another. (See Illus. 98.)

Moulding planes were in use during the same period of time as wooden bench planes. Much of the information you read about bench planes will also apply to these tools.

# Types of Moulding Planes

Moulding planes can be grouped into three general categories: shaping, simple, and complex planes. Of the three, shaping planes—also called hollows and rounds—are the least complicated. Simple moulding planes make just one basic moulding profile, while complex planes make a combination of two or more of the simple shapes.

## Hollows and Rounds

Shaping planes are called hollows and rounds. (See Illus. 99.) These names refer to the shape of the tool's sole rather than the shape the tool makes, a designation that distinguishes shaping planes from simple and complex planes. The sole of a hollow plane is concave, and the sole of a round plane is convex.

Hollows and rounds are very simple tools, but can be difficult to understand. They were made in graduated sets of pairs: each hollow plane had a matching round plane. Tool factories sold them in sets of 9, 10, 18, 20, and 24 pairs. Thus, the smallest set contained 18 planes, and the largest had 48 planes.

Usually, the arc that forms a hollow plane's concave sole (or a round plane's convex sole) is ⅙ the circumference of a circle. This means that the distance across the sole is a chord that is the same length as the arc's radius. More concisely, the distance across the sole is one half the diameter of the circle the plane works. When you measure a pair of these planes, it is easiest to measure point to point on the hollow. So, a hollow plane that measures ½ inch works a one-inch-diameter circle. So does its matching round.

In a set of hollows and rounds, the pairs are usually numbered sequentially. So, the number stamped on the heel is not necessarily the length of its radius. A set of nine pairs is often numbered from 1–9. These pairs of planes are usually graduated in increments of ⅛ inch of radius, with the first pair starting at either ⅛ inch or ¼ inch. If pair #1 is ⅛ inch, each succeeding number *will* correspond to the number of eighths of an inch. For example, #5 will be ⅝ inch and #6 will be ⁶⁄₈ (¾) inch.

If pair #1 starts with ¼ inch, the number on the heel *will not* correspond to the number of eighths. Pair #2 will be ⅜ inch, #5 ¾ inch, and #6 ⅞ inch.

In a set of nine pairs, the planes can also be numbered using even numbers from 2–18. In this case, the number again corresponds to the number

*Illus. 99. Pairs of hollows and rounds. The largest pair, which measures 2 inches in radius and is marked #30, was made by Chapin-Stephens, a company that was formed in 1901.*

of eighths of an inch of radius. For example: pair #2 would usually have a radius of ¼ inch (²∕₈), and the radius of pair #4 would be ½ inch (⁴∕₈), etc., up to #18, which would have a radius of 2¼ inches (¹⁸∕₈ inch).

The largest set of 24 pairs is usually numbered from 1–24. In this case, the number on the heel again corresponds to the number of eighths of an inch of radius. However, to make things even more confusing, this system sometimes changes after #20. The last four planes, #21–#24, increase by quarter-inch increments. Thus, pair #21 has a radius of 2¾ inches, ¼ inch larger than pair #20. Pair #24 has a 3½-inch radius.

If this is not confusing enough, not all American tool companies used these systems. And the English used a different method, so their numbers do not correspond to the same numbers on American planes. Some English hollows and rounds are numbered from 1–20, and increase by increments of ¹∕₁₆ inch, except after #12 when they increase by ⅛ inch. Thus, pair #1 is slightly under ⅛ (listed as ⅛ bare). Pair #2 is a full ⅛ inch, #3 is ³∕₁₆ inch, #4 is ¼ inch, and #5 is ⁵∕₁₆ inch. Pair #12 is ¾ inch, but #13 is ⅞ inch.

Sets of hollows and rounds are not common; most sets have been broken up over the years. If you buy a set, and its numbering does not conform to any of the systems described above, you can probably deduce the numbering from the information you have. Most of the hollows and rounds you will find come as single planes that have been separated from their original sets. I will not buy lone planes, as they are not very useful without their mates. I assembled my set of hollows and

rounds by buying pairs. I ignore the numbers stamped into their heels and instead designate them by the length of their radii, writing the size on the end grain in ink. If you do this, measure point to point on the sole of the hollow.

When you need a plane, you can usually just look at the width of its sole to conclude whether or not it will do the job that has to be done. If you own an assortment of pairs made in both England and America, look up the maker's mark to determine if it is an English or American plane.

Most simple and complex moulding planes have two features, called the fence and stop, built into the sole. These features keep the tool square against the edge that is being moulded and stop the cutting at a certain depth. They also help you to cut identical mouldings every time you use the plane. The fence and stop are described in greater detail on pages 84 and 85.

Hollows and rounds do not have either a fence or a stop. In concept, they are similar to spokeshaves in that they are used for shaping. However, because they are moulding planes, they work on straight edges, rather than curves.

Hollows and rounds are very uncomplicated tools, although they do very complicated work. For example, these planes will make lengths of replacement mouldings. They are handy for making repairs or for making precise copies when you cannot find a moulding plane of the original shape.

It is surprising how often hollows and rounds come in handy for general woodworking. For example, a round plane run against a straight edge clamped to a board will by itself make flutes;

a hollow will make reeds. If you own hollows and rounds and are familiar with how they work, you will discover all kinds of uses for them—jobs you once tried to do with carving tools, scratch tools, a router, etc.

# Simple Moulding Planes

These moulding planes make just one basic profile, but unlike hollows and rounds which do shaping, they are truly moulding planes. Simple moulding planes make simple profiles that include the bead, the cove, the thumbnail, the ovolo, the astragal, and the S-shaped cyma curves (also known as ogees). (See Illus. 100.)

The bead is the most widely known simple moulding. This basic profile has always been, and continues to be, so common that you will certainly recognize it even if you do not know it by name.

A bead is a half round on the edge of a board that's set off by a narrow groove called a quirk. In the past, woodworkers referred to this moulding as a quirked bead. Every craftsman owned at least a couple of bead planes, as well as a couple of hollows and rounds. More advanced joiners probably owned entire sets. A set would have contained many of the following sizes: $\frac{1}{8}$, $\frac{5}{32}$, $\frac{3}{16}$, $\frac{1}{4}$, $\frac{5}{16}$, $\frac{3}{8}$, $\frac{1}{2}$, $\frac{5}{8}$, $\frac{3}{4}$, $\frac{7}{8}$, $\frac{8}{8}$, and $1\frac{1}{4}$ inch.

Bead planes, along with hollows and rounds, are by far the most common moulding planes sold on the second-hand tool market. You may have to sort through anywhere from 10–30 or more bead planes and hollows and rounds to find one plane that makes any of the other simple or complex shapes described below.

The thumbnail is another simple moulding; it's called a thumbnail because its profile looks like the end of your thumb when it's viewed from the side. It is similar to a quarter-round moulding,

only its shape is based on the ellipse, rather than the circle. During the 18th century, it was commonly used around drawer fronts, the lids of chests, and the edges of stiles and rails on raised panel walls and doors. During the 19th century, the bead, the ovolo, and sometimes the ogee often replaced the thumbnail for these uses.

The ovolo is a quarter-round moulding set off on both ends by a short, straight drop called a fillet. An astragal is similar to the ovolo, except that the astragal is a half-round moulding set off by two fillets. The cove is a concave quarter-round moulding.

Some of the most common profiles used during the 18th century were the S-shaped curves called cyma curves. The true S curve (concave at the top and convex at the bottom) is the cyma recta. The reverse S curve (convex at the top and concave at the bottom) is called a cyma reversa. (See Illus. 101.)

During the 19th century, cyma curves were also given the name ogee curves (also referred to as OG). Distinguishing between these names is somewhat complicated because the reverse S-shape (the cyma reversa) is called an ogee, while the S-shape (the cyma recta) is called a reverse ogee.

As with bead planes, moulding planes that made simple profiles came in graduated sets. The size of the moulding plane is usually stamped into the heel. This is generally expressed as a complete fraction (for example, $\frac{3}{4}$). However, it can be expressed in the number of eighths of an inch (for example, #6 = $\frac{3}{4}$ inch).

# Complex Moulding Planes

Complex mouldings are made up of two or more basic shapes. For example, a large cyma recta

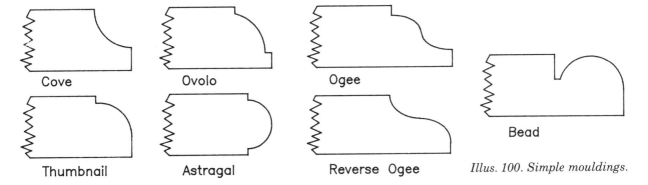

Cove  Ovolo  Ogee  Bead

Thumbnail  Astragal  Reverse Ogee

*Illus. 100. Simple mouldings.*

*Illus. 101. A moulding plane used to make curved mouldings. It makes a cyma reversa (OG) moulding. It was made by Moseley & Son, King St. & Bedford St., Cove Garden, London.*

over a smaller cyma reversa was a very common 18th-century cornice (the large moulding placed at the intersection of the walls and ceiling). In Illus. 102, some other complex mouldings are also shown, and are identified according to some of their common uses.

Over the nearly three centuries that moulding planes were in use, various profiles appeared and disappeared according to fashion. As a result, there are distinct Georgian, Federal Greek Revival, and Gothic shapes. (See Illus. 103.) All these periods occurred simultaneously in England and America. It is possible for someone who is knowledgeable about mouldings and architecture to pick up a complex plane and assign a date as well as a likely use to the mouldings it made.

# Marks

Many moulding planes found in America were made in England. These planes can be difficult to identify because (like bench planes) those made in both countries look alike. Also, the same mould-ing profiles were popular simultaneously in both countries.

Like bench planes, moulding planes often have names stamped into the end grain of the plane's stock. Most are owners' stamps, and these will usually be less elaborate than a maker's mark. You generally have to look at the maker's marks to identify where a plane was made. Lists of English and American moulding plane makers have been compiled, and some are included in some of the books listed in the Bibliography on page 254.

Collectors like moulding planes more than bench planes. Although the prices of average planes are still quite inconsistent, some planes are very valuable because of their maker's marks. For example, during the early 18th century there was a group of plane makers in southeastern Massachusetts in the towns around Rehoboth and Wrentham. This group included such famous plane makers as Francis Nicholson and Cesar Chelor. Some of these planes made by these craftsmen sell for a great deal of money.

*Illus. 102. Complex mouldings. 1 is an astragal and cove, and was commonly used as a chair rail, under stair treads, and on window stools. 2 is an astragal and cyma recta (reverse OG). It was used as a chair rail and to cap hand rails on banisters. 3 is a cyma recta and cove and was used as a cornice.*

Complex 1

Complex 2

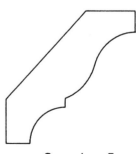

Complex 3

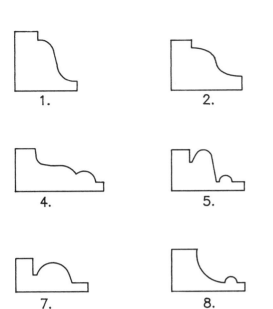

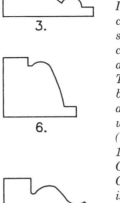

*Illus. 103. Over time, mouldings changed with fashion, and someone who is knowledgeable can assign a moulding an approximate date and likely use. The mouldings shown here are all back bands, the moulding applied around the outside of fireplace, window, and door architraves (casings). Number 1 is a 17th-century back band, 2 is early Georgian, 3 and 4 are late Georgian, 5–8 are Federal, and 9 is Greek Revival.*

# 18th- and 19th-Century Planes

The earliest moulding planes were made in the same craft shops where bench planes were made. In the second quarter of the 19th century, during the Industrial Revolution, the manufacture of moulding planes moved from the craft shop to the factory. Once again, bench planes and moulding planes were made in the same factories.

Just as the sizes and shapes of factory-made bench planes became very standardized, so did the sizes and shapes of moulding planes. Therefore, most 19th century planes look very much alike, and any woodworker who knows what to look for can easily distinguish between what tool dealers refer to as "18th- and 19th-century" planes.

The terms 18th and 19th century are somewhat arbitrary. Eighteenth century really means handmade, while 19th century means factory-made. However, the craft tradition of plane making continued for well into the 19th century. Therefore, planes referred to as 18th-century planes could very well have been made as late as 1825, about the time factories were being built.

The terms 18th century and 19th century apply more to a plane's characteristics than to the actual date it was made. If a dealer's stock of moulding planes were to be arranged side by side

on a shelf or stacked end up in a box, you can quickly pick out the earlier planes. Nineteenth-century factory planes rarely vary in length, and are a standard nine and one half inches long. There is a good possibility that any moulding plane in the collection that is either longer or shorter than the other moulding planes was made earlier.

Even this technique is not absolutely foolproof. Nineteenth-century woodworkers often stored their planes end-up in boxes and would occasionally cut the stock of a longer, earlier plane, to make it fit. This alteration destroys the collector value and antique price of the plane. But, since shortened stock does not necessarily affect the way a tool works, this type of alteration can be a bargain for a woodworker who wants to use the plane.

When you do find a longer 18th-century plane, take it down from the shelf and compare it with later planes. You will notice some other interesting differences between this handmade plane and one made in a factory. (See Illus. 104 and 105.)

A factory that produced as many as a hundred planes a day used machinery. As a result, all of its plane stocks and wedges were standardized. The stock's shoulders are plain, and embellished by only a slight concave channel. Four decorative chips were generally carved from each corner of a factory plane. The edges of 19th-century planes were softened only slightly.

*Illus. 104. A group of "18th-century" planes that were handmade by plane makers working in the craftshop tradition. Note the variety in length and in the shapes of the stocks and wedges.*

*Illus. 105. A group of 19th-century factory-made moulding planes. Even though they were made in different factories, these planes look remarkably alike.*

A craft-shop plane maker sought to make a plane that looked good and worked well. The chamfer that relieves the corners and edges of the stock of an 18th-century plane was done by hand and is very pronounced, a true bevel. The layout lines are often visible. These craftsmen even used moulding planes to decorate the moulding planes they were making, and almost always ran a fully developed moulding on the shoulder. While a cove was the most common moulding used, some makers used an ovolo or a cyma recta.

Earlier planes made in the craft-shop tradition display a great deal of individuality. When individual characteristics of furniture are identified, they can be used to distinguish furniture made in different regions and occasionally by individual cabinetmakers. The same method of identification is now being used on the different characteristics of 18th-century planes. For example, certain plane makers used distinctive wedges that can be used to help identify their work.

Like bench planes, 18th-century planes were made of a variety of hard woods, either the wood

the maker preferred or the wood he had on hand. Nineteenth-century moulding planes were generally made of beech or yellow birch.

## Boxwood Inserts

There are other ways to distinguish 19th-century planes from earlier handmade ones. Plane-making factories developed an innovation that was only rarely used, if at all, during the craft-shop period: Boxwood inserts set into the soles of some planes to prevent wear. The inserts are called "boxing," and such a plane is said to be "boxed." (See Illus. 106.)

A bench plane's sole will wear out eventually, and has to be jointed. However, very little can be done to correct the sole on a moulding plane. When it wears out, the tool is useless. So boxing was inserted at the points in the sole where the most friction (and the resulting wear) will occur. Since boxwood is an extremely dense wood that is more durable than even beech, boxing increased a plane's working lifetime. The boxing could be as simple as a narrow strip placed at just the wear

*Illus. 106. Several different types of boxing. The boxwood inserts wore better than the native hardwood (usually beech) used to make the stock.*

points, or the entire sole could be made from a length of box keyed into the stock with an elaborate joint called a dovetail (although the term dovetail used in this context is different from its meaning in cabinetmaking).

Factories called their boxed planes their "best" planes, and planes without boxing were called "common." Moulding planes with boxing are often in better shape than those without it, so you should be on the lookout for them. Boxed planes should be more expensive, but often they are not.

# Plane Parts and Characteristics

The best way to familiarize yourself with a moulding plane is to examine one. When you do, you'll discover that besides the body and sole, these tools have many other parts in common with bench planes. The sides are called the cheeks. The right-hand cheek is open so that the chip can be ejected. The opening is called the throat. The opening in the sole that leads to the throat is the mouth. The iron rests on the rear surface of the throat, called the bed, while the forward surface is called the wear. The front end is the toe, and the rear the heel.

Remove the iron. Though the wedge's shape looks like it was meant to be driven upwards by striking it just under the top lobe, do not do this. The wedge is made of thin stock, and can be easily damaged. Also, the lobe has a very short length of grain, and a sharp blow can snap it off. As you

examine moulding planes in dealer shops, you will certainly find planes with wedges that have been damaged in this way.

First, remove the wedge. It will lock when it is tapped in place, but eventually this lock will release due to the stock's seasonal movement. Try pulling it out with your hand. If the wedge is not loose enough to be pulled out of its slot, rap the stock on the heel with a wooden mallet or a flat-faced hammer such as a cobbler's hammer. A regular carpenter's hammer can damage the wood.

If the wedge will not loosen after the blow from the hammer, perhaps it is stuck in place by rust or grime. To loosen it, drive the stock off the wedge. Turn the plane upside down and grip the wedge in the jaws of a vise so the wedge is held securely and safely while you drive the upper rear corner of the heel. (See Illus. 107.)

Like the irons of bench planes, moulding plane cutters are often tapered so that a permanent lock is created by two tapers working against each other. If the wedge still will not release, try driving the tang of the iron downwards with a hammer. This will force the blade out of the stock via the mouth. So that this problem does not reoccur, never set the wedge too tightly.

## Pitch

Moulding planes were used in many branches of woodworking, but primarily by cabinetmakers and interior joiners. Although these crafts are similar in some respects, there is one important distinction: the interior joiner did not make the

*Illus. 107. If you cannot remove a moulding plane's wedge either by pulling it with your hand or by rapping on the heel, do not strike the wedge. Grip it in a vise and drive the stock off it.*

same demands on his planes that the cabinetmaker made. He worked almost exclusively with pine, poplar, or other soft conifer and deciduous woods that were selected from the best boards he could find.

On the other hand, cabinetmakers usually used hard woods to build furniture. Moreover, they used hard woods with a pronounced figure, which is difficult to work. Curly and bird's-eye maple will try men's souls, as well as the soles of their planes.

A cabinetmaker's moulding planes often had a high pitch. As discussed in Chapter 9, the steeper bedding angles are called the York, middle, and half pitch. A plane with a common pitch is more likely to have belonged to a craftsman who worked in building construction. (See Illus. 108.)

# Fence and Stop

Almost every moulding plane, except hollows and rounds and some thumbnail planes, has two critical surfaces that regulate its use: the fence and the stop. The fence is on the left-hand side of the sole, and when pressed against the edge of a jointed board it prevents the plane from wandering either left or right. (See Illus. 109.) The fence is also at a right angle to the stop, and while the plane is in motion the fence keeps the tool in the proper position relative to the surface of the work.

The stop is on the right-hand side of the sole. Each time the moulding plane is run down the edge of a board, it removes a shaving and brings the moulding closer to its final form. Each pass also brings the stop closer to the upper surface of the board. Eventually, it makes contact with this

*Illus. 108. Moulding planes have different pitches; those with higher bedding angles worked best on figured woods. Clockwise from the upper left are planes with the following pitches: common pitch, York pitch, middle pitch, and half pitch.*

*Illus. 109. This cove and astragal profile was often used as a back band on Federal period door and window architraves. The vertical detail on the left is the fence. The horizontal detail on the right is the stop.*

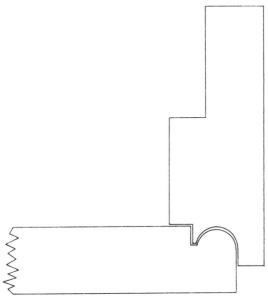

*Illus. 110. Depending on the fence, some moulding planes (such as the bead) are held vertical to edge of the board.*

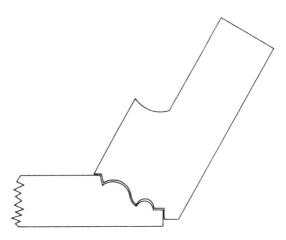

*Illus. 111. Other planes are sprung. Their fences require that they be held at an angle to the edge. Such a plane usually has spring lines that indicate to the user how to hold it correctly.*

surface and rides on it, preventing the iron from engaging the wood. It has stopped the cutting action. During the final pass of the plane, the fence rides on the board's outer edge, and the stop on the upper surface.

# Spring

Because of their fences some moulding planes have to be held vertical to the edge of the board and at a right angle to the upper surface. Bead planes, for example, are held this way. (See Illus 110.)

A vertical position works for simple, shallow mouldings. However, on wider or deeper profiles the plane works more efficiently if it is held at an angle, biased towards the person using it. This position, with the plane essentially pushed downwards at about 45 degrees against the arris (the corner formed by the edge and upper surface), helps to keep the plane on track. (See Illus. 111.) This angle relative to the surface of the work is called the "spring", and a plane that is made this way is said to be sprung.

Many 18th-century planes are not sprung, and can be difficult to use. The left-hand edge has to cut down almost to the finished depth before the rest of the profile is developed. This means that just the fence keeps the plane tracking, without help from the moulding's shape. It is very easy for one of these deep (not sprung) planes to jump off the board. When you find these early planes, you'll notice that they often have been given a second, wider fence by a former owner who decided to correct this problem. (See Illus. 112.)

If you buy a plane that jumps easily off the edge, first make sure the problem is the fence and not a wracked stock. This problem is discussed on

Illus. 112. When a fence is too worn to track on the edge of a board, the only solution is to add an extension. This 18th-century plane makes a moulding that was often used in the Georgian period to make the cap for a stair banister.

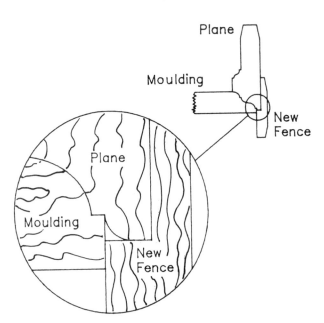

Illus. 113. Some early planes have very shallow fences. When the fence becomes worn, it will not track well, so an extension has to be added to it.

pages 87–90 under the head Selecting and Reconditioning a Plane. If, however, you discover that it is the fence, you can make an extension. (See Illus. 113.) Rabbet a piece of hard wood to the same depth as the fence's thickness. Set the rabbet over the outside edge of the plane and attach it with wood screws. The surface that extends below the old worn fence should be flush with the original. This new extension will keep the plane on track. Also, it damages the plane as little as possible. Usually, three small screw holes are all that are needed.

A sprung plane usually has some guides from which you can tell how to hold the tool. On the plane's toe, you will see two (sometimes just one) scribed lines. These are called spring lines, and they intersect to form a right angle. Each line is a continuation of the fence and/or the stop. Since these two surfaces are at 90 degrees to each other, so are the two spring lines. When there is only one line, it usually is a continuation of the fence

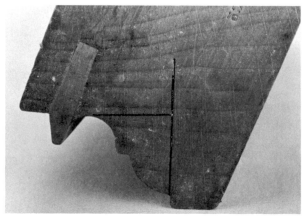

Illus. 114. The spring lines on this moulding plane have been darkened so that they can be seen in the photograph. One line indicates the vertical, and the other the horizontal. When the spring lines are parallel to the top and the edge of the board, the plane is in position to be used.

and tells you how the plane should be held relative to the vertical.

If you tilt the plane until the spring lines are parallel to both the board's edge and upper surface, you automatically bring the plane to the angle at which it is supposed to be used. Before beginning each cut, just glance downwards to make a quick final check of your angle. You can also keep an eye on this angle while the plane is in motion. (See Illus. 114.)

If you have a sprung plane that does not have spring lines, you can make them yourself. Before buying a plane, make sure that the lines relate to the profile the plane makes. If not, the plane has perhaps been damaged or altered.

# Selecting and Reconditioning a Plane

If you want to use moulding planes in your woodwork, you will quickly find purposes for them. The more planes you acquire, the more versatile you will become. It is still possible to quickly and inexpensively assemble a graduation of hollows and rounds and bead planes. You will have more difficulty with ogee, ovolo, cove, and thumbnail planes, but these are by no means rare.

Gathering a working collection of complex planes that has any reasonable continuity is even more work. (See Illus. 115.) Fewer of these planes were made than simple planes, and some profiles and sizes are scarce. However, unless you are doing restoration work where it is necessary to make exact duplicates of missing mouldings, you do not need continuity in your collection. Several complex shapes of different widths will be ample for most woodworkers.

Although the second-hand tool market is becoming more sophisticated and efficient, pricing is still inconsistent. Specialized dealers know what is scarce and desirable to collectors, but much of the time moulding planes are priced "by the inch." This is a facetious way of saying that a bigger plane is more expensive than a smaller plane, regardless of what profile the tool makes, its condition, and rarity.

Like all second-hand planes, moulding planes can be found in unexpected places and at unusual prices. I have a friend who was dining in a restaurant one night when he spotted a large moulding plane hanging on the wall. It was there, along with other old tools, as part of the decor. The plane makes a complex profile that was commonly used as a Georgian chair rail. My friend asked to see the manager. He explained that he used such tools in his work and offered to buy the one on the wall or replace it. Since the restaurateur was only concerned with the effect the plane gave his restaurant, he agreed to swap. My friend now owns and uses the tool, and the wall of the restaurant sports a well-worn jack plane.

Many planes offered for sale on the second-hand market as moulding planes are not. Be aware of these planes. Most are special-purpose planes whose bodies look like that of a moulding plane. Chapter 13 discusses these special-purpose planes.

As with all other second-hand tools, moulding planes were inherited by people who did not appreciate them. As a result, most that have survived are in bad condition. These planes are precision instruments, much more so than even bench planes. A moulding plane's action is so complex and the tolerances so tight that defects are almost impossible to correct. So, unless you've come across a moulding plane that is rare or that's needed so badly that you cannot wait to find another, only buy those that are in almost perfect condition. A defective plane that makes a common profile is just not worth reconditioning.

Make sure that the stock is sound. Minor check cracks are not a problem. However, if the plane has dry rot, extensive bug damage, or severe cracks it should be rejected.

The stocks of some moulding planes look worse than they really are. Although still sound, they are just badly dried-out. I have done wonders for such planes by standing them on one end in a coffee can that's filled about one third with boiled linseed oil. The dried wooden stock absorbs the oil. In fact, it's surprising how much oil the wood absorbs.

Reverse the stock periodically, so that the other end is submerged. Repeat this process for up to half a day. The stock will look much better, and you'll notice that some check cracks have even closed. Wipe the plane dry and set it aside in a warm area. Small amounts of oil may bleed back out. Of course, always dispose of rags soaked with linseed oil. They can combust spontaneously.

If a bench plane is missing its iron, you can give it a replacement with only a minimal amount of work. A moulding plane without its cutter, however, is a serious problem. It is not that the work is that difficult. Rather, the problem for the inexperienced is in conceptualizing how the cutter and its bezel work. And, when you do understand these concepts, grinding a new cutter is time consuming.

Unless the profile made by the plane is rare, or you need it immediately, it is seldom worthwhile to cut a moulding plane iron. Still, if you find you enjoy working with moulding planes, chances are that you may someday encounter this problem.

I pick up stray irons whenever I find them, but there are problems with moulding plane irons that do not occur with bench plane irons. The web (the wide part) of a moulding plane cutter is ground to the same profile as the sole. If the profile is deep, the contour will reach well back into the steel (like bench plane irons, these blades often have a forge-welded layer of steel). There

simply may not be enough space remaining to grind a new, or very different, shape.

If you have this trouble, either find a blacksmith who will make a new cutter out of tool steel or make one yourself. The instructions for heat-treating a new iron are given in Chapter 6. Later, I will describe how new cutters are ground.

Before purchasing a tool, check to see if the plane's stock is warped or wracked. Hold the plane up to your eyes and sight down the fence from toe to heel. Make sure that it is straight enough to track on the edge of a board. If the plane body is no longer straight, do not bother to buy the plane. It will never work correctly again without being almost totally remade.

Before buying a plane, take a moment to check the sole. Unlike bench planes, moulding planes cannot be resoled with much success. If this surface is worn or battered, it is not worth reconditioning. Only minor corrections can be made to a sole, and you must know how to do them before determining whether or not they are worth doing.

All the details (especially the fence) should still be crisp. Otherwise, as the plane makes each successive cut, it will have difficulty tracking. If the plane rides off its cut, it will probably damage the moulding.

Check as to whether the fence or stop have been tampered with by a previous owner. If the plane has been recently reworked, the old, darker surface has been disturbed, exposing wood that is lighter in color. However, if the alteration was done long ago, by now the freshly exposed wood has had time to age.

You will also find planes that have been given a new profile. If the tool was damaged, a former owner may have changed the sole. If the plane makes a shape you want and still works well, buy it. But you will not be able to determine whether or not it will work until you have first become familiar with moulding planes. Until you are, only buy those that are in good (and original) condition.

Make sure that the wedge on the plane is the original one. If a tool is missing its wedge, a dealer will often fit it with a spare. If this replacement wedge does not fit properly, it will not hold the iron tightly. Although you can make a proper wedge, it will be time consuming. If the plane does not have its original wedge, you have a good reason for asking the dealer to lower the price.

The passage and ejection of the chip is a very complicated process, and if it is not accomplished precisely the plane will choke. While the wedge is in place, check its lower end by looking in the throat. The end should be pointed and not blunt. The wedge must also fit perfectly against the wear (the front surface of the throat). If it is not tight and a gap has formed between the pointed end and the wear, the shaving might jam into this opening. Every time you use the plane it will choke. So if the wedge does not fit perfectly, you will have to make a new one.

The plane will also choke if it has an iron with a badly worn profile. The high points on the cutter cut the lowest points on the moulding. They cut almost all the time the plane is in operation. The low points on the cutter cut the high points on the moulding. They only cut during the final passes.

The high spots on the iron receive the most wear. Eventually friction takes its toll. To make these worn areas engage the wood, you have to set the iron so deeply that the unworn, low areas make too heavy a cut. This chokes the plane. (See Illus. 116.)

Beware of alterations to the throat before buying the plane. It was common for later owners who did not understand the tool to think that choking occurred because of a problem in the plane's throat, when in reality it was the iron. They often tried to correct the choking by opening the throat so that the chip could clear more easily. (See Illus. 117.) This created more difficulties, as the altered throat could no longer eject the chip.

The boxwood used to box the soles of best-grade 19th-century planes was harder than the native hardwood used to make the stock. As a result, it usually took longer to wear out. Unfortunately, boxwood is more brittle, and can be easily damaged if it is bumped or dropped.

The boxing was often set into simple grooves, and if damaged, it can be removed with a chisel and a new strip can be laid in. This procedure is much more difficult if the boxing is held with the complicated keyed slot (called a dovetail). If dovetail-keyed boxing is damaged, do not purchase the plane. It would be a major undertaking to remove the old boxing and make a replacement.

Once you have used moulding planes and have become familiar with them, you can do some work that may be necessary. Earlier on, I cautioned you not to strip an old finish. Still, there are times when this has to be done. Should you be forced to do so, make sure that as little stripper as possible flows into the throat. Carefully clean any stripper out of this passage by twisting a soft, dry rag and pulling it through the wedge slot. Give the stock and wedge a light coat of linseed oil that's cut 5–1 with turpentine. Allow plenty of time for drying before reassembling the plane.

Occasionally, you may find a plane that is missing its wedge. More often, the wedge has been badly damaged by a former owner who thought it was supposed to be removed by driving up on the lobe with a hammer. Although a miss-

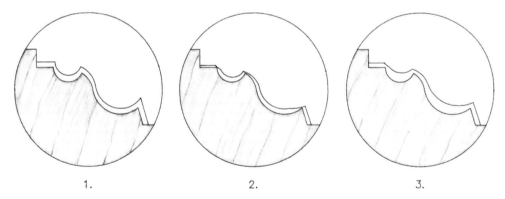

1.  2.  3.

*Illus. 116. 1. When you sight down the sole of a moulding plane with a properly sharpened and adjusted cutter, you will see a halo of steel rising above the sole. 2. Over time, friction wears down the edge's high points and they become too low to engage the wood. The plane takes a few shavings and then stops cutting. 3. The temptation is to advance the cutter to raise the worn spots. When you do, the low spots take such a heavy cut that they choke the plane. The only solution is to reshape the edge.*

*Illus. 117. Do not buy a moulding plane with an altered throat.*

ing or damaged wedge will greatly diminish the plane's value to collectors, it is nothing more than a minor annoyance for a woodworker.

When making a replacement wedge, use the same species of wood as the stock. First measure the wedge slot (or what remains of the original) and plane a piece of hardwood to that thickness. It is critical that the wedge taper at the same angle as the bed and wear, or else it will not work. If the wedge's taper is greater, it will only make contact at the top. (See Illus. 118.) If the taper is less, the wedge will only bind at the bottom. (See Illus. 119.) At best, a wedge without the correct taper will not hold. At worst, it could damage the stock.

If part of the old wedge remains with the plane, copy the angle from this. You will have to do some final fitting and some fine trimming to make sure that the wedge fits properly and locks along its entire length. I use a block plane (see Chapter 5) for this type of trimming.

Once the wedge fits snugly in its slot, you have to bevel the end on the throat side. (See Illus. 120.) This bevel creates a point similar to that on a sharpened pencil (but only on one side). The pointed end prevents the chip from getting caught in the throat and choking it. The bevel also helps direct the chip out of the open throat.

The bevel on the end of the wedge should be a continuation of the incline cut into the stock at the top of the throat. (See Illus. 121.) Once the wedge has been fitted, reach into the throat with a fine scribe and scratch a mark on the side of the

wedge where it disappears into the stock. Bevel from the mark down to the end.

When the wedge's taper and point are completed, finish by shaping the top. There was no sense in doing this first as, had you not fitted the taper correctly, it would have been wasted work.

# Sharpening the Iron

A recently purchased moulding plane almost always has to be sharpened before it can be used. Most woodworkers have difficulty sharpening a straight bezel on a tool like a chisel, and the complex bezel of a moulding plane can be daunting.

To sharpen a moulding plane bezel properly, you first must understand the concept of clearance. Clearance is the relief gained by grinding the bezel. It is what allows the cutting edge to make contact. If an edge does not have sufficient clearance, the excess metal on the heel of the bezel will project down below the cutting edge and will rub on the wood. (See Illus. 122.) It acts as a stop and prevents the edge from cutting any further.

The lower the bedding angle of the blade, the more clearance is needed. Of course if the bedding angle is very low, for example, a block plane at 11 degrees (see Chapter 14), the bezel is placed upwards. If it were placed down, the angle of the bezel would have to be so steep to provide sufficient clearance that the edge would have no strength.

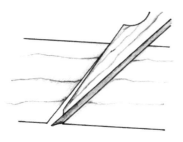 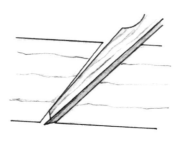 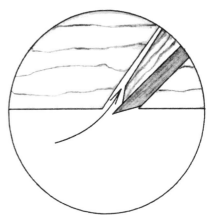

*Illus. 118 (above left and middle). It is critical that a moulding plane's wedge have the correct taper so that it fits tightly in the wedge slot. If it is at too steep an angle, it will only make contact at the top (left). If it is at too shallow an angle, it will bind at the bottom (middle). Illus. 119 (above right). If there is a gap between the end of the wedge and the wear, the shaving will often jam into this opening, causing the plane to choke.*

*Illus. 120. The pointed end on a moulding plane wedge guides the chip up and out of the throat. Most wedges have darkened to three different shades of color. The darker upper end is exposed above the stock. The light area in the middle is where the wedge fits into the stock. The slightly darker lower end is exposed in the throat.*

*Illus. 121. The lower end of the wedge is visible when you look into the throat. Notice the bevelled point, as well as the bevel cut into top of the throat. These features help guide the chip out of the throat. Note also how tightly the wedge is fitted, so as to eliminate any gap where the chip could become snagged.*

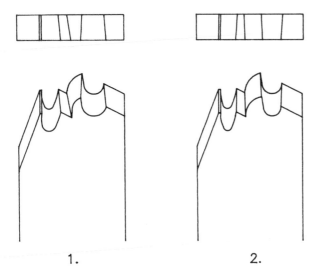

Illus. 122. The bezel of a moulding plane has to have proper relief. 1 is a drawing of a very elaborate Federal period back band. It has six different cutting surfaces. Above it is an end view, showing how the bezel would appear if you held the cutter up to your eyes, and sighted along its length. It shows the angles at which the individual features are relieved. 2 shows how the same cutter would appear if the various details were not given any relief. The plane would take several passes, but would soon "stall" as the unrelieved areas began to drag on the wood.

When you have to make a new blade for a moulding plane, consider each individual segment of the cutting edge (whether curved or straight) as a separate cutter, complete with its own bezel. Remember, each fillet (the short, straight sections between separate shapes) has to have its own clearance as well. (See Illus. 123.)

If you have to make a new iron for a moulding plane, begin by finding a spare cutter that will fit. This may be as difficult as anything else you will do during this process.

You may be forced to make a new blank out of tool steel. If so, first anneal it so that the steel is as soft as possible. Next, cut the blank wide enough to make the web and long enough so that the tang extends all the way up the throat and projects beyond the end of the wedge. This is usually between 7 and 9 inches. The web does not have to be any wider than the side-to-side width of the mouth, from the inside of the closed cheek to the outside of the open one.

Make sure that the web's top edge does not press tightly against the top of the throat. There

Illus. 123. Each segment of the profile on a moulding plane cutter has to have its own separate relief.

has to be enough room here for the cutter to be adjusted longitudinally. Leave about a quarter inch of space for adjustment.

Cut the tang so that it is slightly narrower than the wedge. If it is too tight, it will be difficult to fit it up the wedge slot and there will not be any room for lateral adjustment.

It is not essential that the blank be tapered, but (if you have the equipment to do this) a tapered blank will fit more tightly with the taper of the wedge.

Once you have made a suitable blank, spray the forward side of the web with layout fluid and allow it to dry. Place the blank in the stock and slide in the wedge. Put some pressure on the wedge to lightly secure it. Adjust the blank so that it is bedded in the same position as when used. Trace the sole's outline in the layout fluid with a sharp point (I use a scratch awl with a long shank) held as parallel to the sole as possible. (See Illus. 124 and 125.)

Remove the iron and grind to that shape. You may not be able to make the entire profile on a grindstone; if not, do the rest of the work with files. Chain-saw files, which are available in several smaller diameters, are handy for this work. Triangular files, used for sharpening hand saws, are also useful. You can also cut the profile in the web with a jeweler's saw (which looks like a fine coping saw) and clean up with files.

Place the iron back in the stock and check the profile against the sole. As you sight down the sole, you should see a perfect "halo" of metal that uniformly follows the sole's profile. Take the time

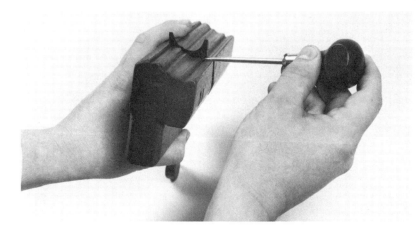

*Illus. 124. To make a replacement cutter for a moulding plane, first spray the forward surface of the annealed web with layout fluid. Then put the cutter into the plane and hold it in place with the wedge. Trace the shape of the sole with a scratch awl. This plane makes a profile known as an astragal and cove. The cutter I am using for a replacement once belonged to a nosing plane.*

*Illus. 125. Look closely and you can see the outline of the moulding traced in the layout fluid. Grind and file to this line.*

*Illus. 126. A moulding plane with its cutter properly adjusted. If you sight down the sole towards the mouth, you should see the edge as a uniform halo above the sole. Every point on the edge should be equidistant from the wood.*

to place the iron back in the stock correctly; otherwise, all your work will be wasted.

Remove the iron and spray the back surface of the cutter with layout fluid. Return the iron to the stock and secure it with the wedge. Drive the tang until the forward edge is projecting about ⅛ inch. This setting would be too heavy for use, but is necessary to lay out the bezel.

With the scratch awl, trace the moulding profile in the rear surface of the cutter, back ¼ inch from the edge. This new line will outline your bezel and relief. It is not perfect, but it is accurate enough to be used as a guide.

Now, grind and file the bezel. Make sure that each individual shape has its own separate relief so that it has the proper clearance.

When you are done, put the iron back in the plane and adjust it so that the cutting edge just barely projects above the sole. Adjust it as you would for use. Now look down the sole from the toe towards the mouth. You should see a perfect "halo" above the sole. (See Illus. 126.)

Turn the plane around; you should also see a halo as you look from the heel towards the mouth. If you have insufficient relief anywhere along the bezel, you will be able to spot it from this direction. Look for any area on the bezel that might project below the cutting edge and make contact with the surface of the moulding.

If you are making your replacement cutter from an old iron, follow essentially the same process. Begin by annealing the iron to make the steel soft and easier to work. Do not file the tempered blade even if you can because you risk working back into softer areas of the steel when you make the new profile. These spots will not hold an edge, and will wear faster than the other areas. The extra effort needed to heat-treat is worthwhile if it avoids this problem.

Once again, spray the front edge with layout fluid and set the wedge so that none of the old profile intersects with the new one. Then, trace the profile and make the new shape. After that, the process is the same as described above.

Whether you made your own replacement cutter or reshaped an old one, you next have to harden and temper it. Use the techniques explained in Chapter 6.

By now, you should understand why I suggest not buying moulding planes with common profiles if they are missing their irons. Making a new one

can take several hours, and the value of your time usually exceeds the value of the plane.

A missing iron is not the only problem you may discover. Even if the iron is the original one, it may be worn and will still require considerable work before the tool will operate properly. Remember, as a moulding plane is used the high points of the cutting edge receive more wear than do the low points.

If the high points are considerably more worn than the low ones, you will have to rework the profile. Remove the cutter from the stock and anneal it. Once again, spray the forward surface with layout fluid. Once this coat of fluid is dry, return the iron and wedge to the stock. Adjust the iron so that the worn spots are level with the sole of the plane, with the unworn areas projecting. Run the point of the awl along the contour of the sole in front of the mouth. The scribe will trace a line in the colored fluid that describes the restored profile.

Remove the iron from the plane and clamp it, bezel-up in a vise. When you restore the original profile, you will reduce the width of the bezel and lose some clearance. You will have to reestablish the old bezel. If you do not, there is a risk of insufficient relief. After doing this, follow the process as described for making a new cutter.

If the dealer has several planes of a particular profile that you want to use, check the irons. All other considerations being equal, buy the plane with the least-worn iron. Before making the purchase, set the iron as you would when using the plane, and sight down from the toe. If you see that the halo of metal projecting above the sole is not equidistant everywhere, some areas perhaps not showing at all, you will have to do work on that cutter.

Whether you have made a new iron, reworked an old one, or just restored the original profile, you now have to harden and temper the cutter as described in Chapter 6.

Finally, hone the cutter. Begin by polishing the surface opposite the bezel. (See Illus. 127.) This usually requires a trip to the lapping table. Be sure to flatten just the web. Like bench planes, moulding plane irons are often signed by their maker. This mark adds to the plane's value, especially if it is a good, clean impression. You reduce the resale price of the tool if you scratch or obliterate the mark on the lapping table.

*Illus. 127. Before honing the bezel of a moulding plane cutter, flatten it on a lapping table and polish it on a honing stone.*

You will need a variety of slipstones to sharpen the bezel. I use ceramic files that are made of the same material as my sharpening stones. Use them dry; it is less messy this way, and there is no oil to obscure your view when you look to see how the edge is developing. The set of files include a round, a triangle, a square, and a teardrop. The triangle and square files are good for corners and fillets as well as convex edges. The round and teardrop files work on concave areas.

Slipstones are sometimes sold on the second-hand market, often in the same places where you find old tools. They are seldom recognized or appreciated by dealers, and are often priced well below the cost of a new one. Otherwise, you can buy slipstones from any new tool dealer or through a tool catalogue.

You may have to strop the cutter to get rid of any wire edges. I use a soft buffing wheel impregnated with buffing compound. A stiff pad will not reach down into the contours of a moulding plane iron unless you exert a lot of pressure. This pressure can wear away the high points before the low points are sufficiently stropped. Do as little stropping as possible. When you have to, do it as quickly and lightly as you can to avoid wearing the high spots.

There are many planes now available on the second-hand market that were made and used in England for many years by English woodworkers. In the last 20 years, they have been swept up in large numbers by American tool dealers and shipped to the United States. England is more humid than my native New England. As a result, the wooden bodies of planes that have been shipped here will experience some dimensional change as the stock balances its moisture content with the new relative humidity. Usually, the stock becomes slightly narrower.

Because the plane's cutter is made of metal, it retains its original size. The shrinkage causes the features of the sole to move slightly out of alignment with the same features on the cutting edge. If you try to use the plane, it will act very much the same as it would were the cutter's profile badly worn. It will begin to form the moulding, but the cutting will soon stop when the sole reaches the first point of misalignment.

Because their stocks are thicker and experience a greater amount of shrinkage, this problem is more acute on wider moulding planes than narrow ones. I recently purchased a plane that had this problem. The stock was 2¼ inches thick. The total misalignment amounts to approximately ³⁄₆₄ inch, enough to prevent the plane from working.

The stock's dimensional change is most noticeable on planes with a complex profile, especially one that has deep features such as a quirk. A wide round plane, for example, would experience a similar amount of shrinkage, but probably not enough misalignment to make the tool unworkable.

You can easily spot this misalignment when you know what you are looking for. (See Illus. 128.) Sight down the sole, from the toe towards the heel; you will see that the features appear to shift right to left in relation to those on the iron. This is not really a right-to-left movement; it only appears that way. Actually, the whole stock has shrunk uniformly. Because the iron (which re-

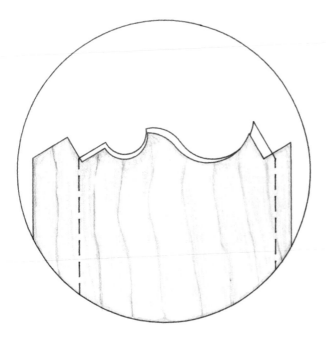

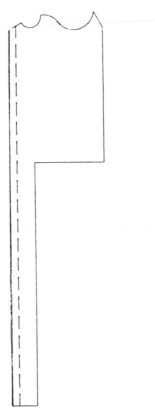

*Illus. 128. If a wooden moulding plane is transported from a humid climate to a much drier one, the stock will shrink. As a result, the cutter's features will no longer be aligned with the same features in the sole. The sole will appear to have shifted to the left. As you sight down the sole, it will look as shown here. The dotted lines indicate the edges of the cutter's web.*

*Illus. 129. The solution to the misalignment caused by shrinkage is to file the left edge of the cutter's web and tang. This allows you to center the cutter on the sole's profile.*

tains its original width) is butted tightly up against the closed cheek, it cannot be centered with the features in the sole's profile.

I have only experienced the problem of dimensional change on English planes that have been recently imported. I assume that the problem also occurred with planes that were imported a century or more ago, but was corrected by the original owners. The same dimensional change could probably occur with American planes transported from the American eastcoast (or mid-west) to arid regions like the American southwest. The amount of dimensional change in English planes imported to the American Southwest must be even greater than in the planes I have come across.

Happily, the solution is simple. Draw-file the left edge of the iron, the edge that butts against the closed cheek. (See Illus. 129.) Concentrate on the web and the lower length of the tang. Remove about the same amount of metal as the displacement caused by the shrinkage.

Now, the iron can be centered on the sole. The effects of the shrinkage are distributed evenly over the whole profile and often become so slight as to no longer affect the tool's performance. If not, you will have to follow the procedure already described for restoring the profile.

After sharpening the cutter, return it to the plane, invert the stock and, holding the iron so that the bezel faces the toe, insert the tang up the throat and through the wedge slot. Turn the plane upside right with the sole down and keep the iron from falling out by covering the mouth with your index finger.

Insert the wedge and push it in place snugly enough to hold the iron. Once again, sight along the sole from the toe to the mouth. Lightly tap the tang end of the cutter with a hammer until you see the cutting edge rise slightly above the sole. If necessary, adjust the iron laterally by tapping either side of the tang. (See Illus. 130.)

When you are done, the cutting edge should produce a halo above the profile of the sole. It should be equidistant everywhere. If you have set

*Illus. 130. Adjust a moulding plane by sighting down the sole. To advance the iron longitudinally, tap the end of the tang. Retract the iron by tapping the heel of the stock. To laterally adjust the iron, tap the sides of the tang.*

the iron too deeply and have to retract it, tap the heel. Once you have reached the desired setting, tighten the wedge with the hammer. Remember, someday you will once again have to loosen the wedge, so don't be too aggressive.

# Using a Plane

Your moulding plane should now be ready to use. Select a board. Most moulding planes were not meant to be used on either very hard or heavily figured woods. The exceptions are those with a high pitch—for example those with a half or middle pitch. The board should be made of pine or some other soft, even-textured wood with a straight grain. Plane the upper surface with a smooth plane and joint the left edge.

Clamp the board to the workbench with the left edge projecting slightly over the front side. Set a marking gauge (see Chapter 18) to the width of the finished moulding and use it to scribe a line parallel to the edge of the board. Set the plane's fence against the board. The mouth should be just off the right-hand corner, so that your first pass cuts a shaving from the entire length of the edge. (See Illus. 131.)

If the plane is sprung, bring the stock to the correct angle relative to the vertical and horizontal. Remember, you do this by looking at the scribe lines on the toe.

Grip the plane by the toe with your left hand and by the heel with your right. Your right palm, pushing against the upper corner of the heel, will provide most of the force. Make sure that your hands are not covering the mouth or are otherwise interfering with the chip. Remember, the chip does not exit through the top of the stock as on a bench plane. The throat is open on the right-hand side and the chip is ejected onto the workbench.

As you walk alongside the board, push the plane forward while simultaneously pushing down on it to prevent it from jumping off the edge. Do not stop the forward motion, as this will create a bump in the moulding's surface. However, if the tool does stall, no great damage has been done because this bump will be removed in the next pass.

Repeat this process, and with each pass of the plane you will see the moulding begin to develop. (See Illus. 132.) As the plane works down deeper into the profile, the chips will become wider and the plane will become harder and harder to move. You will have to lower your torso to lean into the work.

The amount of force required to move a moulding plane depends on the width of the moulding. This can be deceptive, since the real width of the cut is the width of the moulding's surface—in other words, its length if all its curves and fillets were straightened out. A narrow, but deep moulding will cut a wide chip and require a lot of force to move it forward.

When the profile is complete, the stop will ride on the right side of the moulding and lightly burnish the board where it rubs. It is very important that the plane not stall during the last pass. (See Illus. 133.)

A thumbnail moulding plane is a special case because some of these tools have no fence and

*Illus. 131. Before starting the first pass, make sure that the spring lines are aligned with the edge and surface of the board. Note the marking gauge line. It has been darkened with a pencil so that it would show in the photograph.*

*Illus. 132. After you have made several passes with the plane, the moulding will begin to develop. Note the shavings, which also get wider with each pass.*

*Illus. 133. The moulding is complete when the stop rides on the board.*

instead have two stops, one on either side of the sole. (See Illus. 134.) To begin cutting with a thumbnail plane, hold the middle of its concave sole on the arris of a jointed board. These planes are usually sprung, so the plane is also at an angle.

Rather than establishing a track that the plane can follow, each pass of the thumbnail rounds the arris to the plane's elliptical profile. (See Illus. 135.) The curved edge grows wider until near the very end, when the fillet is cut into the board's surface. When the fillet is fully formed, the right-hand stop rides on the board's upper surface. The left-hand stop is the area of the sole below the edge of the cutter. When this runs on the edge of the board below the curve, it stops the plane from advancing laterally, to the right. (See Illus. 136.)

Once you have finished the moulding, cut it from the board using either a ripsaw or a table saw. Remember to follow the line made by the marking gauge. Joint the outside edge to remove the saw marks. To plane the sawn edge of a large moulding, clamp it in the side vise of a workbench and run a plane over it. Short lengths of moulding can be jointed on a shooting board. Very small mouldings can be jointed if you invert a jointer plane and hold it in the vise. Pull or push the moulding over the mouth of the plane.

It is not likely that you will ever have to make mouldings with hollows and rounds (remember, this is not all these planes are used for). If you do, you will find it helpful to have another plane called a V-plow. This tool is a special-purpose plane and is similar to the grooving plow plane described in Chapter 13.

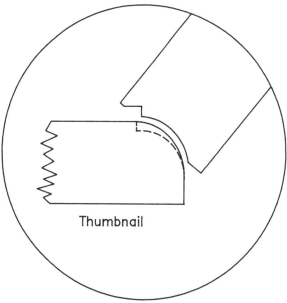

Thumbnail

*Illus. 135. Start a cut with a thumbnail moulding plane on the board's arris; each pass will round the corner more and more. Eventually, the drop (indicated by the dotted line) will form. Shortly after that, the stop will come into contact with the board's surface.*

The V-plow plane has a movable fence and depth stop, so it is adjustable for both width (distance from the edge) and depth. It is used to establish the deepest points on the moulding profile you are making. (See Illus. 137.) When you are done with this tool, the rough form of the moulding will exist as one or more V grooves of different depths that are spaced across the edge of the board. Hollows and rounds are used to shape the remaining profile. Rounds hollow out the concave areas and hollows shape the convex.

V-plow planes are rare, and your chances of finding one are not good. However, if you own a combination plane such as the Stanley #45 (see Chapter 15), you can use it for this purpose. You will have to make a V cutter, which is a simple process.

Making a moulding with hollows and rounds is more time consuming than using a moulding plane and works best if you are making a limited amount of moulding. However, at times I have had to make a large quantity of moulding, and have found it more economical to make a moulding plane of that profile. To do this, first purchase a single round plane that has become separated from its hollow plane. Give the round plane a new sole and grind its iron to the new profile as described above. The result is a moulding plane with the profile that you need. I only recommend doing this in an emergency, as the process takes time. Also, to do it properly requires a complete understanding of how moulding planes work. This can only be learned through experience.

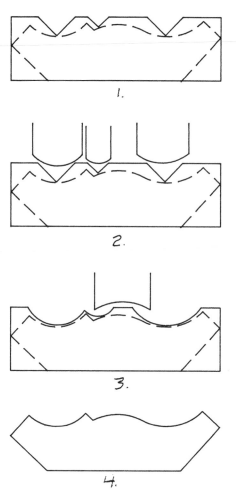

*Illus. 136. When making a thumbnail profile, start in the middle of the curved profile. You will gradually round the corner of the board. Eventually the drop will form and the stop will make contact with the board's surface.*

*Illus. 137. Mouldings can also be made with hollows and rounds. Shown here is a cyma reversa and cove cornice. As in 1, first lay out the placement of the various details and their depths with a V-plow. I use a Stanley #45 for this purpose. As shown in 2, use various widths of rounds to excavate the concave details. Use hollows of various widths to shape the convex details. As shown in 4, the fillet between the cove and the cyma recta can be trimmed with a shoulder rabbet, and the corners with a bench plane. (I am grateful to Malcolm MacGregor of Durham, New Hampshire for this information. Malcolm, who specializes in handmade mouldings for restorations and reproduction buildings of all periods, frequently makes special-order mouldings with hollows and rounds.)*

# 13
# Wooden Special-Purpose Planes

Many of the specialized shaping functions that woodworkers now perform with table saws and routers can also be done by hand with specialized planes that can be found on the second-hand tool market. Many of these functions include general carpentry, joinery, and cabinetmaking jobs such as making grooves, rabbets, dados, and tongue-and-groove joints. Other planes do more specific jobs such as making window sash, raised panels, and the rule joint on the leaves of drop-leaf tables. Special-purpose planes also do some shaping functions, such as smoothing concave surfaces, as well as nosing and rounding.

The history of these planes closely parallels those of bench and moulding planes. During the craft-shop period, special-purpose planes were made by plane makers who worked predominantly by hand in the craft-shop tradition. These early planes can be highly individualistic. During the second quarter of the 19th century, special-purpose planes were manufactured in plane factories. As a result, they became standardized. Like bench and moulding planes, factory-made special-purpose planes look very much alike.

Until the third quarter of the 19th century, almost all special-purpose planes were made with wooden bodies that resemble those of wooden bench and moulding planes. Much of the same terminology is even used to describe them. As a result, a lot of the information and advice given about bench and moulding planes will apply to special-purpose planes.

Metal plane-making technology improved during the last quarter of the 19th century, and as a result some wooden special-purpose planes disappeared altogether, while others were replaced by a metal version. During the mid-20th century, small portable machines, such as the router, were developed and used for many of the special functions once performed by wooden special-purpose planes. Because wooden special-purpose planes were made for almost three centuries, some can be more easily found than their later, metal equivalents. This does not mean that metal planes are rare, but rather that some of the wooden versions are quite common.

If you choose to incorporate second-hand tools into your woodworking you will discover that, like moulding planes, special-purpose planes can be easier to use than the machines generally relied on to do the same job. In many cases they are less time-consuming, especially in a small shop where most jobs are done one at a time, rather than in multiples.

## Adjustable Plow Planes

Grooves are used in a lot of joinery and cabinetmaking, like, for example, in stile and rail door construction, raised panel walls, furniture carcasses, drawers, and feather-edge sheathing. A groove is either run on the outside edge of a board (as in door construction) or on the surface of the board, close to the edge (as in drawer construction). Grooves are commonly confused with dadoes. To avoid this mixup, just remember that a groove runs with the grain of the board, while a dado runs across it.

In hand woodworking, the adjustable plow plane makes grooves. A plow plane can be adjusted in three ways. It allows you to make grooves of various widths, depths, and distances from the edge.

Best-grade plow planes are often very elaborate; they were made of exotic woods such as rosewood or box, and decorated with ivory and brass. (See Illus. 138.) Collectors compete intensely for them, and they are among the most expensive tools on the second-hand market. However, good serviceable plow planes are still available for woodworkers at affordable prices. These common-grade planes were made of beech or other native hard woods.

Some plow planes have a closed tote that is an integral part of the stock. In other words, the stock and tote are made out of a single block of wood. Razee bench planes also have a closed tote, but this is usually a separate piece, joined to the stock. Other plow planes were made without a tote, and in this case they have rectangular bodies similar to moulding planes.

Instead of a true sole, a thin, two-part metal guide is attached to the bottom of a plow plane body. This guide is called either a keel or a skate. It is, of course, narrower than the narrowest cutter that would be used in the plane.

The front part of the keel is mounted in front of the iron. Its back end is relieved and bevelled to form a rudimentary throat and to throw the chip to the right (onto the bench). The other piece of the keel is mounted behind the cutter, which rests on its forward edge. (See Illus. 139.) The keel stiffens the iron and helps it to resist chatter. Thus, its purpose is like that of the bed in a moulding plane.

## Width of Groove

The width of the groove the plow plane will cut is determined by the width of the iron. English and American wooden plow planes were originally sold with a set of eight irons, graduated by ¹⁄₁₆ths of an inch. (See Illus. 140.) The smallest is usually

*Illus. 138. A group of plow planes and a set of irons. The plow plane on the far left and the one that's third from left are often called "Yankee" plows. Some planemakers made these as "common"-grade planes. You can adjust the fence on the first plane by loosening the wedges. To adjust the wedge on the second plane, you would have to loosen a pair of thumbscrews on the top of the stock. The plane on the far right is made of solid boxwood. Its manufacturer would have sold it as a "best"-quality plane.*

*Illus. 139. Plow plane cutters are heavily tapered; this helps them to resist chatter. They are also supported by the front edge of the rear keel. This surface is usually bevelled to produce a narrow edge that fits in a groove cut into the rear of the cutter. This prevents the cutter from moving side to side.*

*Illus. 140. American and English plow plane irons are graduated by ¹/₁₆ths of an inch, starting at ¹/₈ inch (²/₁₆). They are numbered sequentially; the number is often stamped on the upper end of the cutter. Thus, a #2 iron is ³/₁₆ inch, #4 ⁵/₁₆ inch, and #5 ⁶/₁₆ (³/₈) inch.*

¹/₈ inch, and the widest ⁵/₈th inch. They are numbered sequentially rather than by actual size. Therefore, the ¹/₈-inch iron is #1, the ³/₁₆-inch iron #2, up to #7, which is ¹/₂ inch. Iron #8 then jumps to ⁵/₈ inch.

Plow plane irons have a distinct shape. The tang is thin and quite wide, about ⁵/₈ inch. The upper end is often round with a hooked corner. The maker's mark is usually stamped into this flat surface of the tang, as is the number that designates the width.

From the side, a plow plane iron looks like a mortise chisel. The lower end is strongly tapered front to back, up to ⁵/₁₆ inch thick just above the bezel. This thickness is necessary to strengthen the blade, as it is only bolstered by the keel, not by a true bed as in a moulding plane.

# Depth of Groove

The groove's depth is regulated by an adjustable stop that's set into the stock just in front of the iron. (See Illus. 141.) On some common-grade plow planes, this stop is no more than a sliding wooden bar. It is inserted through a vertical slot in the stock and held in position by a wooden thumbscrew. In humid weather, when the stock swells, this simple type of stop can become bound in its slot.

On other plow planes, the stop is made of cast iron and is adjusted by a brass thumbscrew. The screw has right-hand threads, so the sole is advanced by turning the screw counterclockwise. This type of stop is easier to adjust and is not affected by humidity.

# Placement of Groove

The groove's placement (its distance from a parallel edge) is also adjustable, and is regulated by means of a movable fence that slides on two arms that pierce the stock (See Illus. 142.) On some plow planes, the arms are threaded; each arm has two wooden nuts. The inside nuts hold the fence a set distance from the cutter, while the two outside nuts lock the fence in place by pressing it against the inside nuts.

On other plow planes, the arms are locked in place by wedges, which are more difficult to adjust accurately. Since the fence has to be perfectly parallel to the metal keel, the wooden nuts were a definite improvement.

Thus, by virtue of its set of irons, depth stop, and movable fence, the typical adjustable plow plane can make a groove from ¹/₈ to ⁵/₈ inch wide that's up to one inch deep and up to 4¹/₂ inches from the edge of the board.

# Selecting a Plane

As with all old tools, people who inherited adjustable plow planes did not always treat them well. The wooden threads are delicate and can be easily chipped. Damage to the threaded arms or missing nuts is one of plow planes' most common flaws. Small nicks in the threads are not necessarily a problem, but if they are too badly damaged the nuts will slip and the adjustable fence will not hold its setting.

Do not buy a plow plane that is missing any of

*Illus. 141. The depth of the cut made by a wooden plow plane is controlled by the depth gauge, a movable stop (often metal) between the keel and the fence. On this plow plane, the depth gauge is moved by means of a brass thumbscrew on the top of the stock.*

*Illus. 142. The distance between the groove and the edge of the board is controlled by a movable fence. On many plow planes, this adjustment is made with two wooden nuts. You can adjust some plow planes (often called "Yankee" plow planes) by loosening the wooden wedges.*

its nuts. Making new ones will require finding a tap and die set that will make the same threads that are on the arm. (A tap and die is a set of threading tools. The die makes the screw, and the tap the corresponding nut. Taps and dies are made for working in either metal or wood.)

A plow plane will work without a complete set of irons, but you will be limited as to the widths of grooves you can make. I pick up spare irons whenever I see them. Tool dealers do the same, and as a result they often have assembled sets of plow plane irons for sale. Although these irons may not have originated with your plow plane, and may even have been made by different com-

panies, they almost always fit. An assembled set will be much less expensive than a set by a single maker.

## Sharpening the Iron

A plow plane's cutters have a straight cutting edge ground at a right angle. Grinding it square is very easy because the edge on the widest cutter is only ⅝ inch. Remember, the bezel is not hollow-ground; it is straight or even convex, like the bezel on a mortise chisel (see Chapter 17).

Next, polish the side opposite the bezel, a process that is very easy since this surface is so narrow. Finally, hone the bezel. If it is convex, you cannot ride on the heel and edge as you would with a chisel or bench plane iron. You just have to be careful to not hold the cutter at too high an angle. This will round the cutting edge, creating a burr.

## Using a Plane

The first step in using a plow plane is to select a cutter. The one you'll select depends on the groove's width. Remove the iron that is in the plane. It is held in place by a wooden wedge that is shaped like those in moulding planes. Do not drive up on this wedge since it can be easily damaged. If the plane has a tote, you cannot rap the heel. If the wedge cannot be pulled out by hand, remove the iron by driving it out of the mouth. (See Illus. 143.) Do this by tapping the tang with a hammer. The iron has such a pronounced taper that it loosens easily.

Place the desired cutter in the plane, as you would a moulding plane iron. Make sure that the cutter is centered on the keel by tapping on the sides of the tang. Adjust it longitudinally by tapping the tang's rounded upper end.

Set the depth stop to height and adjust the fence for placement. Clamp the board in place, either edge or face up, depending on where the groove is to be run. (See Illus. 144.) In stile, rail, and panel construction the groove is on the edge of the stiles and rails, and these would be clamped edge-up in a side vise. On a drawer side, the part would be clamped on the edge of the bench top.

*Illus. 143. If you cannot pull a plow wedge loose with your hand, drive the cutter out the mouth. It is heavily tapered, and the lock will release very quickly. Pull the iron out the rest of the way with your fingers.*

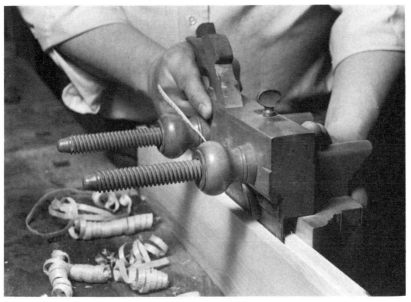

*Illus. 144. Using a plow plane. Note the curled chips.*

Press the fence against the left side of the wood and place the cutter against the rear edge. Push the plane steadily, and "walk" it along the board. The chip—a very distinctive one that curls like old-fashioned bologna curls—ejects out of the right side, as on a moulding plane. Make each successive pass in the same way until the cutting action is stopped by the adjustable sole.

# Dado Planes

Dadoes differ from grooves in that they run across the grain rather than with it. They can be made with a special type of wooden plane. Unlike the adjustable plow plane, a dado plane has only one iron and can cut a dado of just one size. Like plow plane irons, these irons are graduated in sixteenths of an inch, usually starting at ¼ inch and ranging to one inch.

Dado planes look very much like moulding planes and are often mistaken as such by less-knowledgeable dealers. However, there are some important differences that will help you to spot one very quickly. (See Illus. 145.) The first is the scribing iron in the front of the stock. This iron, whose purpose will be discussed later, is held in place by a wedge that is a smaller version of the ones used to hold moulding plane irons. However, it is vertical (or nearly vertical) to the stock rather than at an angle to it.

Although dado planes cannot be adjusted for width, they can be adjusted for depth. Set into the stock is a depth stop that is very similar to those found on adjustable plow planes. Like adjustable plow planes, some dado planes have an adjustable wooden post secured by a wooden thumbscrew. Others have a cast-iron (sometimes cast-brass) sole that is adjusted by either a setscrew in the side of the stock or a thumbscrew on the top.

Stamped into the heel among the owners' and maker's marks is the plane's size (the width of the dado it makes). This is usually expressed as a fraction, for example, ½ ⅞, ⅝, etc. On other dado

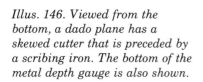

*Illus. 145. A group of dado planes. Note the shape of their throats. The plane on the right (³/₈ inch, by W. Sergent) is of "best" quality. It has a metal depth gauge that is adjusted by a brass thumbscrew on the top of the stock. The upper left plane (³/₁₆ inch, by M. Copeland) is "common" grade. It has a wooden depth stop held in place by a wooden thumbscrew. The lower left plane (⅞ inch, unmarked) is also "common" grade. Its depth stop is made of brass and is held in place by a metal screw.*

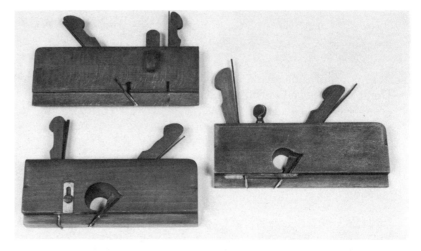

*Illus. 146. Viewed from the bottom, a dado plane has a skewed cutter that is preceded by a scribing iron. The bottom of the metal depth gauge is also shown.*

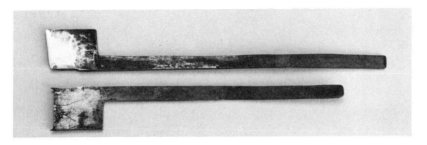

planes, the size is designated by a single number, as it is on plow plane irons. Thus, #1 is ¼ inch, #2 ⁵⁄₁₆ inch, and #3 is ⅜ inch, etc.

The sole of a dado plane is set off from the stock by a rabbet so that the sole is the width of the dado it makes. The mouth runs the entire width, and is open on both sides. (See Illus. 146.) The depth gauge allows a maximum depth of usually ½ to ⅝ inch.

There is yet another difference between dado and moulding planes. Since dadoes are cut across grain, a dado plane's cutter is set askew. The skewed edge helps it to lift the chip without tearing out the edge of the dado. To set the iron askew, the bed is also angled, as is the wear. Because the iron is set at two angles, its pitch and the angle of skew, the wedge has to be slightly different than that on a moulding plane. Its cross section is a parallelogram. (See Illus. 147.)

The iron is skewed so that its right-hand edge is leading. This means that it will throw the chip to the left. Because the chip is cut across the grain, it is very different from a chip made by a moulding plane. Instead of being peeled off in continuous sheets, the chip tends to crumble and break into short sections.

Because a dado plane cuts a different chip than a moulding plane, the throat is different. Both cheeks are open, but not the same amount. The throat hole is tapered so that it forms a section of a cone. Because the iron pushes the chip to the left, the throat is more open on the left-hand side of the plane, and the chip is ejected from this side.

A dado plane iron also has a more pronounced taper than does a moulding plane iron. A moulding plane's throat is closed on the left side, which (along with the bed) supports the cutter. Since the dado plane is open on both sides, the cutter has to be heavier to stiffen it and reduce the chatter.

To further prevent tear-out as it cuts across the grain, the dado plane has a scribing iron mounted in front of the cutter. This means that when the tool is pushed across a board, the wood ahead of the cutter has already been scored by the scribing iron. All the iron has to do is cut the waste out of the dado.

The scoring is done by two teeth whose outside edges line up with the edges of the sole. The scribe is set into its own vertical (or nearly vertical) slot and is held by the smaller wedge already described. The scribing iron projects from its own narrow mouth which, like that of the cutter, is open on both sides of the sole. Because the scribing iron does not lift a chip, it does not need a throat.

## Selecting and Reconditioning a Plane

Dado planes were used for one specific purpose: to make dadoes. They were not given the same amount of use as were bench and moulding planes, and, as a result, they are generally in better shape than these tools. Since dado planes are not rare, do not buy one that has been abused or has any major flaws.

Dado planes have more parts than bench and moulding planes, and are more likely to be missing one or more of these parts. Also, examine the parts to make sure that they belong to that plane, and are not replacements.

You can easily remake a wedge for either the cutter or the scribe. The wedge for the cutter is the more difficult of the two, since its front and rear edges are not square to its sides. This means you will have to do more fitting (using a block plane), as all four sides must exactly conform to the sides of the wedge slot.

If necessary, you can also make a new scribing iron. File a blank that's the width of the sole out of tool steel. Do not copy the width of the cutter. Because the cutter is set askew, it is slightly wider than the sole. The scribe iron is not tapered, as is the cutter.

The scoring teeth are sharpened with a knife edge. Gently round the outside edges so that they are convex. Do the same to the bezel (the inside edge) of each tooth. This surface is also convex. (See Illus. 148 and 149.) As a result, the cross section of each tooth is shaped like a football, forming a knife-like edge along its entire outline.

Making a missing cutter is more difficult because it has such a strong taper. If you cannot do this yourself, you might have to have a blank made by a blacksmith or at a machine shop. Remember that the iron is wider than the sole.

You can also make a missing wooden depth gauge. However, don't buy a plane that is missing its wooden thumbscrew, as you need a die that will cut that specific thread. Don't buy a plane that is missing a metal depth stop. The cost of making a new one would far exceed the value of the plane. Finally, do not buy a dado plane that has been altered.

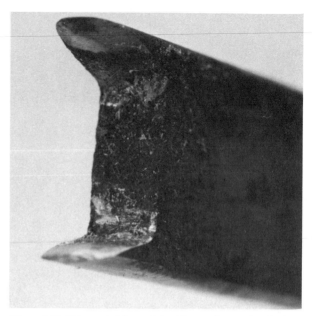

Illus. 149. The outside and inside surfaces of a dado plane's scribes are rounded, the inside more than the outside.

# Sharpening the Iron

Remove the wedge the same way you would remove one on a moulding plane. Never hammer

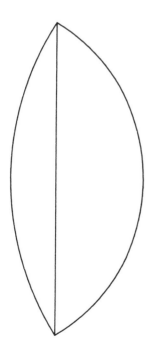

Illus. 148. The cross section of the tooth resembles that of a lopsided football.

the wedge. Gripping the wedge in a vise or driving the iron out of the mouth is usually effective.

Because the iron is set askew, the cutting edge of a dado plane is not square to the sides. If the edge is not given the correct angle, one corner will project lower than the other. Then, to make the low side cut you would have to set the high side so deep that it chokes the plane. And, even if you can get the plane to work, the bottom of the dado will not be square to the sides. The piece fitted into the dado will not butt squarely with the bottom, and a slight gap will occur in the joint.

If you need to regrind a dado cutter, spray the front surface with layout fluid. Set the cutter back in the plane and adjust it so that the lower corner projects above the sole. The higher corner will project even higher. With a fine point such as that on a scratch awl, trace the angle of the sole. Remove the iron and grind to this line. Next, follow the general information on sharpening given in Chapter 8, remembering to first flatten and polish the side opposite the bezel.

Adjust the cutter the same way you would adjust the cutter on a moulding plane. Hold the cutter in the plane with your fingertips while placing the wedge in its slot. Make sure that the

*Illus. 150. Use a wooden dado. Notice the tightly curled chips. I have lowered my left hand so that the plane would show in the photograph. Normally, it would be gripping the toe.*

edge is parallel to the sole. The cutter's sides must line up with the edges of the throat. Advance the setting by tapping on the tang. Retract it by tapping the heel.

You should sharpen the teeth of the scribing iron. It is possible that these have been either broken or worn away. If so, lengthen the teeth (or form new ones) by filing away the metal between them. Remember, the outside edges of the scribe are convex, so the inside bezel of the teeth cannot be flat. They, too, need to be slightly convex. This creates a sharp knife-like edge along the entire profile of the tooth, so that it scores more cleanly and effectively.

Never file or grind the outside edges of the scribe, as this will make it narrower than the iron. It will score a track that is narrower than the dado, and increase the chances of tear-out.

## Using a Plane

The first step is to set the depth gauge. The dado plane does not have a built-in fence. It is used by clamping a stationary fence at a right angle across the work and running the plane against this guide. I usually use a narrow strip of wood. (See Illus. 150.)

To make a dado, first reach across the board you are dadoing. Place the right side of the plane against the strip, and then draw the plane rearwards. This gives the scribe an opportunity to lay out the new dado. Then begin the dado, once again holding the right side against the fence and pushing the plane forward.

If you are doing fine woodworking, you may want a finer finish on the dado's edge. You can clean it with a side rabbet, a tool that is described on pages 113–115.

# Universal Fillister

The rabbet is a third type of cut you often have to make in joinery. A rabbet is cut into the edge of a board; it is open on one side, and has a shoulder on the other. A rabbet cut can be made either with the grain or across it. (See Illus. 151.)

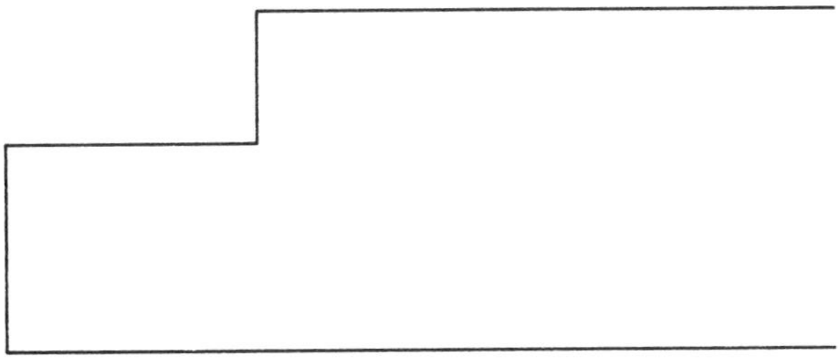

*Illus. 151. A rabbet. Rabbets can be made on either the edges or ends of boards, and as a result run either with the grain or across it. The vertical side is called the shoulder.*

The special-purpose plane that makes rabbets is called a universal fillister. It is also sometimes called a "moving" or "adjustable" fillister since it can make rabbets that vary in depth and in width.

Fillisters made in the craft-shop tradition vary widely. Those that were factory-made are standardized, and their stocks look like those of large moulding planes. They are usually made of beech. (See Illus. 152.)

Because rabbets vary both in depth and width, fillisters share some similarities with plow and dado planes. The first similarity is the adjustable depth stop, mounted on the side of the stock.

*Illus. 152. The upper fillister was made by G. W. Denison & Co. of Winthrop, Connecticut. The lower fillister was made by W. Greenslade of Bristol, England.*

On some models, the depth stop is adjusted by a brass thumbscrew. On others, it is held by a metal screw that must be loosened and tightened with a screwdriver every time the depth adjustment is changed. Some fillisters even have a wooden stop and wooden thumbscrew.

A fillister's fence is not outrigged on two arms like the fence on a plow plane, but instead slides back and forth across the plane's sole. Each time it is adjusted, it exposes more or less of the cutting edge. The fence has two recessed slots cut into it, and is held in place against the sole with two metal screws whose heads fit into the recesses. To adjust the fence, loosen the screws. (See Illus. 153.) When you have made the desired adjustment, retighten the screws.

Rabbets can be made either with the grain or across it. So, a fillister's cutter, like the cutter on a dado plane, is set askew, with the right-hand cor-

ner leading. As on the dado plane, the skewed iron is preceded by a scribe that severs the wood fibres ahead of the cutter and helps avoid tear-out.

Instead of a double tooth, the fillister has only one scribe which scores the inside edge of the rabbet. This scribe can be mounted in several different ways. On some fillisters, the scribe is mounted in a small slot in front of the iron and is held by a small wedge. This wedge is often made of box or even brass. (See Illus. 154.)

This type of fillister scribe has a tang and an offset tooth that will be perhaps several times larger than those on the dado plane. The outside of the tooth is rounded, as is the inside bezel, to produce a knife-like edge.

The other type of scribe that can be found on a fillister is held in place by friction. The scribe is slightly tapered; it's narrow at the bottom and wider at the top. The sides of the scribe are also bevelled so that it has a wider back than front, and thus, its cross section is a dovetail. It fits into a groove in the side of the stock that is also tapered from top to bottom, and has dovetailed sides. This tapering in four directions (from top to bottom and front to back) holds the iron snugly in place, and there is no need for either a wedge or a screw.

Tap the scribe with a hammer to make it snug. You cannot tap the sharpened end to withdraw it without causing damage to its cutting edge. There is a horizontal groove cut in the scribe's exposed face. Place a screwdriver bit in this slot and drive it with the palm of your hand. The slightest movement will loosen the friction fit.

A fillister's mouth is more than half the width of the sole, so most of these planes can cut a rabbet up to 1¼ inch. The throat is shaped like that on a moulding plane and ejects the chip out of the right side.

On some factory-made fillisters, the stock's lower right edge is inlaid with a strip of box. This corner receives a lot of wear and the box keeps it from rounding with use.

The iron is heavily tapered, even more than that of a dado plane. (See Illus. 155.) The heavy, lower end helps reduce chatter, although in a wide rabbet some chatter will still occur.

# Selecting a Plane

Fillisters are very common and you should not

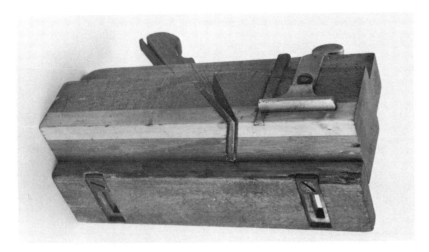

*Illus. 153. You can adjust a fillister's fence by loosening two screws. As it is moved, the fence exposes a greater length of the cutting edge, resulting in a wider rabbet. Note that the mouth is skewed. This is typical of wooden planes meant to be used across the grain.*

*Illus. 154. These fillisters have two different types of scribes. The scribe on the plane on the right is held in place by a wedge made of boxwood. Its teeth are offset from the tang. The scribe on the plane on the left is fitted into a tapered slot. Note the short, horizontal slot; this is where you place a screwdriver bit when loosening the scribe. The depth gauges are also different. You adjust the one on the right by loosening a screw. The other has a brass thumbscrew on the top of the stock.*

*Illus. 155. Because the plane's mouth is skewed, a fillister's cutter is not rectangular. The cutter is thicker than a moulding plane iron; this helps to reduce vibration.*

have trouble finding one. Do not buy a fillister that is not in good condition. Make sure that it is not missing any parts, and also check that all of its parts are the original ones.

## Sharpening the Iron

Like a dado plane cutter, a fillister's iron has a skewed edge.

The angle of its cutting edge is critical if you

want the fillister to cut squarely. If you have to grind the iron, follow the same steps as with a dado plane. Hone the same way.

# Using a Plane

Before using a fillister, set the fence and depth stop to the width and depth of the rabbet you need to make. Make sure that the edge of the fence is perfectly parallel to the edge of the stock before tightening the setscrews.

Set the fillister on the edge of the corner of the board. Hold the plane so that its stock is at a right angle to the board's surface. Push the plane, walking the length of the board if necessary. Repeat the process until the stop ceases the cutting action.

While using the fillister, never forget to hold the plane perfectly vertical. If not, it may roll out of square. Should this happen, you can correct it by holding the fence tightly against the side of the board and at a right angle to the upper surface. While you are making this correction, the iron may not cut the entire width of the rabbet. Several correcting cuts may be necessary before it again cuts a full-width shaving. (See Illus. 156.)

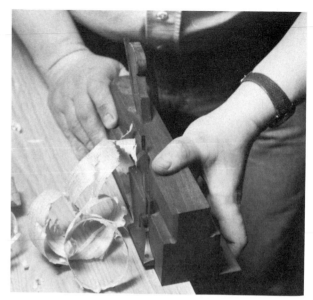
*Illus. 157. Using a fillister to make a rabbet with the grain.*

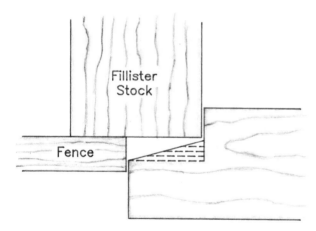
*Illus. 156. If your fillister should work itself out of square, you can correct the problem. Hold the right side of the stock securely against the rabbet's shoulder. Take several correcting passes (as indicated by the dotted lines) until the corner once again forms a right angle.*

When the fillister is run with the grain, it produces bologna-curl chips that fascinate children. (See Illus. 157.) I once made furniture in public at a museum, and whenever I used a

*Illus. 158. Using a fillister to make a rabbet across the grain. Note that the chip breaker has a tendency to make tight curls. These curls are very fragile and break up easily.*

fillister children would always clamor for some shavings to take home with them. When the fillister is run across grain, it makes chips like those made by the dado plane. (See Illus. 158.) These chips tend to break up very easily.

*Illus. 159. A pair of side rabbets. One side is the reverse of the other so that both sides of this unusual type of plane can be shown.*

The fillister may leave the rabbet's shoulder slightly ragged, as this surface is repeatedly scored by the plane's scribe. The bottom surface will often have chatter marks. If you are doing architectural work, such as fitting shelves in a closet, the fillister will produce a suitable rabbet. However, if you are doing finer work, you will want to clean up the rabbet's shoulder and bottom with a side rabbet plane, a skew rabbet plane, or a shoulder plane described in Chapter 14.

# Side Rabbet Planes

Fillisters and dado planes make rabbets and dados that are neat enough for architectural work and for some furniture. However, in fine furniture dadoes and rabbets have to be very clean. Also, when assembling a joint you may occasionally have to make slight adjustments to the shoulders. The wooden special-purpose planes that do this work are called side rabbets. They look very much like narrow moulding planes and are occasionally sold as such by less knowledgeable dealers.

Like table and match planes, side rabbet planes come in pairs. However, in this case there is both a right- and a left-hand plane, and they come in one size. To prevent the planes from becoming separated, dealers will often hold them together with an elastic band.

You can easily spot side rabbet planes on a shelf with other planes, as their irons and wedges are vertical, or nearly vertical. The lowest angle I have measured on these planes was 75 degrees.

Side rabbet planes differ from all the other planes discussed in this chapter. The mouth of a side rabbet plane is on the side of the plane where the throat would normally be. (See Illus. 159.) It is on the right side of a right-hand plane, and on the left side of a left-hand plane. The mouth-side of the stock is perfectly flat and, technically, its lower edge is the plane's sole. The throat is exactly opposite the mouth. In fact, if you hold the plane up at eye level, you can see through the mouth and throat. The iron is set askew, usually at about 65 degrees to the side of the stock. This sets the cutting edge at a high scraping angle.

## Selecting a Plane

Side rabbet planes are not common tools, but they are not rare. If you need a pair, write or call a specialized tool dealer. They often have them in stock.

These tools have a very specialized function: to trim rabbets and dadoes. This is very delicate work, and it is not something past owners did every day. So, side rabbets did not usually receive a lot of wear, and an average pair is in much better condition than an average moulding plane. Do not accept a pair that is damaged or battered. It will be worth your while to wait for a perfect pair; you will not have to wait long.

# Sharpening the Iron

Remove the iron by rapping the heel with a mallet. If this will not loosen the wedge, grip it in a vise and drive the stock off of it. This is easy to do, since the wedge is at (or near) a right angle.

The iron is different from other irons discussed in this chapter because its bezel is ground on the side. (See Illus. 160.) You should not need to reform the cutting edge, as side rabbets were not heavily used. The bezel is steep, around 57 degrees. Begin as you would to sharpen any other cutter, by flattening and polishing the side opposite the bezel. Hone the bezel by holding the cutter on its side. You will have to grip it with just the tips of your fingers. (See Illus. 161.)

Do not buy side rabbet planes that are in less-than-perfect condition. This includes those missing a cutter. However, if you do, or if you inherit a pair that has sentimental value, you

*Illus. 161. The bezel on a side rabbet cutter is vertical, so hold the cutter in the position shown when honing. Grip it with just the tips of your fingers.*

may have to make a replacement iron. Start with a narrow strip of tool steel about ⅛ inch thick. You will have to cut or file the tang, but this will not be difficult while the steel is still soft. If the other iron is still with its plane, use it as a guide, but remember that it is one of a pair and is the reverse of the one you are making.

You can grind the bezel, but it is so steep that you might find it easier to use a file. The edge opposite the cutting edge is tapered so that its forward corner does not project out of the throat. To determine this angle of taper, spray the forward surface with layout fluid, put the iron back in the plane, and trace along the edge of the throat. You can bevel this corner with a file, as well. Finally, heat treat your new iron. Remember to temper the cutting edge, which is along the side of the cutter, rather than the lower end.

If you have to replace a missing or damaged wedge, use the other one as a pattern. Side rabbet wedges are the same shape as wedges on moulding planes. However, since the shaving passes through a side rabbet rather than up and out of a vertical throat, the wedge is shorter and blunter.

Remember that the iron is set askew, so the front and back edges of the wedges are bevelled. Since the planes are opposites, on each plane these bevels are reversed.

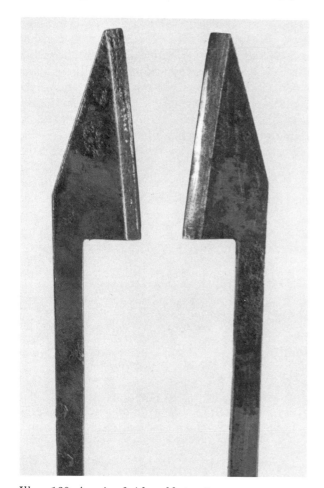

*Illus. 160. A pair of side rabbet cutters.*

## Using a Plane

Return the iron to the plane. The cutter's bottom end must be perfectly flush with the bottom edge of the stock (the surface that on any other plane would be the sole). If it is not, the tool will create a ridge in the corner of the dado where the side and bottom intersect.

The plane should have a very light set. The cutting edge must project out the mouth and be parallel to the side of the stock that acts as the sole. Hold the plane up to your eyes, sighting down the sole. You should see the cutting edge. It should not be projecting more at the top or bottom. (See Illus. 162.) If it does, it will not make perfectly vertical shoulders.

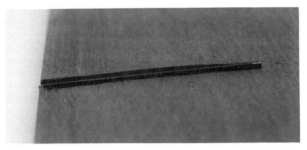

*Illus. 162. The cutting edge of a side rabbet. Strictly speaking, the side of the stock is considered the sole.*

Each pair of side rabbet planes consists of a left and right plane so that you can follow the grain's direction. You do not have to use the right-hand plane on just the right-hand shoulder.

Place the plane in a dado, rabbet, or groove and press it against the surface that has to be trimmed. As you push the plane, it will trim the shoulder. (See Illus. 163.) When used across the

*Illus. 163. Using a side rabbet to trim the shoulders of a dado. The plane pares dust from the shoulder's end grain.*

grain (in a dado or rabbet), the plane will scrape a fine dust-like shaving and leave behind a crisp, finished surface. In a groove or rabbet that runs with the grain, the plane will cut a shaving. The shaving will fall out of the throat, on the opposite side of the stock.

Test the dado's fit. If it is still too tight anywhere along its length, use the plane to lightly scrape the tight spot until it no longer binds.

# Skew Rabbet Planes

The moving fillister is an adjustable plane that's used for cutting rabbets. A similar tool, without the fence and depth stop, is called a rabbet plane, or, more commonly, a skew rabbet plane.

Wooden skew rabbet planes are available in various widths from ¼ up to 2 inches, and a plane's size is usually stamped into the heel. (See Illus. 164.) Skew rabbet planes are not precision tools, and are most useful to carpenters and other woodworkers involved in building construction.

A stock on a skew rabbet plane is very basic, no more than a rectangular block of wood. Its edges are rounded just enough to protect your hands from splinters. The heel's upper corner is also rounded, as this is where you place the palm of your right hand when you use the plane.

Skew rabbet planes have many features in common with dado planes and fillisters. Their cutters are skewed and, along with the mouth, extend the width of the sole, as on a dado plane. The throat is also cone-shaped, and the chip is ejected out of the left side. Unlike that on a dado plane, the sole is the same width as the stock.

Skew rabbet planes sometimes (but not always) have a scribe mounted on the right side of the stock, just ahead of the cutter. This helps when they are used across the grain. These scribes are usually tapered with a dovetail cross section, like those found on some fillisters.

The wedge slot is centered in the stock. Since the iron is skewed, the wedge slot is also angled, as are the front and back edges of the wedge.

## Selecting and Reconditioning a Plane

Skew rabbet planes are common tools on the second-hand tool market, and since they are of little interest to collectors they are generally

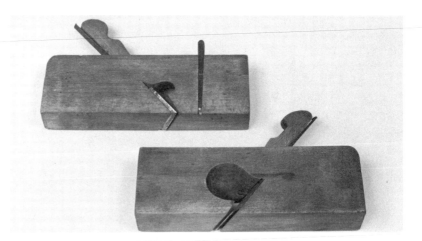

*Illus. 164. The side of one of the skew rabbets shown here is the reverse of the other; this way, both sides of the planes' conical throats can be shown. The upper plane is 1½ inches wide and was made by J. Kellogg, Amherst, Massachusetts. The lower plane is 2 inches wide and was made by Allen Eldredge, "So WmsTown." It may be English.*

inexpensive. Since these tools were used in construction, they are often heavily worn. To work well, the edges of the plane's sole must be crisp, and the mouth should be tight. So when buying a skew rabbet, be selective and accept only those that are close to their original condition.

Before using a skew rabbet, first joint its sole. Do this in the same manner as for a bench plane. I clamp my jointer plane upside down in a vise and run the plane over the jointer.

Use a try square to make sure that the sole is at a right angle to the sides of the stock. If not, you will have difficulty using the tool to square the shoulders of tenons and rabbets.

In time, you may have to inlet a patch into the sole to close the mouth. If the mouth is not tight on a skew rabbet you have bought, make this repair before using the tool. Follow the steps described in Chapter 9.

Wooden skew rabbet planes are satisfactory for heavy work, but the wooden sole does not create a mouth that is tight enough for fine work. To maintain tighter tolerances, some plane makers attached a metal sole. This created on a wooden plane the narrow mouth that is necessary when you are trimming rabbet and tenon shoulders to closer tolerances.

You, too, can apply a metal sole. Use wood screws and countersink the holes in the metal strips so that the screw heads are slightly below the surface of the new sole. Joint the sole on the lapping table.

## Sharpening the Iron

Like fillister and dado cutters, a skew rabbet's cutter is heavily tapered. The cutter differs in

that the tang is centered above the web. (See Illus. 165.) It should be sharpened in the same way as dado plane and fillister cutters. It is essential that the cutting edge is parallel to the sole. If not, follow the same process used for regrinding a dado plane cutter.

Many skew rabbet planes are missing their scribes. You can make a new one out of tool steel, using files to do the shaping. These scribes can also sometimes be found in the "dollar box" in second-hand tool shops.

*Illus. 165. The cutters from the two skew rabbets are parallelograms. The tangs are centered above the web.*

## Using a Plane

Skew rabbet planes are handy for smoothing or adjusting a tenon's cheeks and shoulders. (See Illus. 166.) They will also clean up a large rabbet that has been made by a moving fillister. When

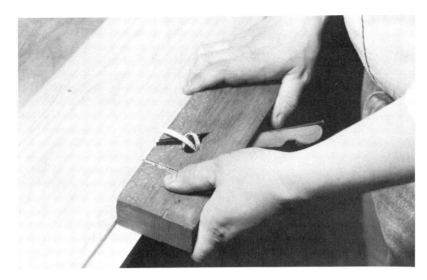

you need to make a rabbet that is wider than your fillister, you can use a side rabbet. Clamp a wooden strip parallel to the board you are working, and use this as a stationary fence.

# Match Planes

It is the nature of wood to continually balance its own moisture content with the relative humidity of the surrounding air. On a humid day, wood absorbs moisture from the air, and on a dry day moisture is drawn out of the wood. As a result, wood is continually in motion as its width increases and decreases relative to its moisture content. This means that in dry weather or during the cold periods when heat is used, gaps will open in any surface made up of boards that are butted together, such as a sheated wall, a floor, or the backboards on a piece of case furniture.

Woodworkers use many methods of construction that allow for expansion and contraction. Perhaps the two best known are stile, rail, and panel construction and tongue-and-groove construction. Today, tongue-and-groove boards (often called matched boards) are available in lumberyards in standard dimensions. If you need something more specialized, you can make your own tongue-and-groove boards with a special pair of wooden planes called match planes.

Best-grade, factory-made match planes have a closed tote like the one found on plow planes. The tote and stock are made out of a single block of wood, usually beech. Common-grade match planes have a stock that looks very much like that on a moulding plane. (See Illus. 167.)

Some factories produced an inventive alternative: a two-in-one match plane that made both the tongue and the groove. The tool had the tongue cutter on one side, and the groove cutter on the other. You first use one side of the plane, and then turn it around. I find these two-in-one planes awkward and uncomfortable to use, and, as a woodworker, recommend buying a pair of match planes. Leave the two-in-one version for tool collectors.

Since match plane cutters are the opposite shape of the profile they make, it is easy for the beginner to confuse the two planes. The groove plane has a cutter that looks like the tongue, while the tongue cutter is notched in the middle, so it looks like the groove. (See Illus. 168.)

Unlike plow planes and fillisters, match planes are not adjustable, and each pair is intended for work on boards of only one thickness. Versatility is obtained by owning several different-sized pairs. Match planes were usually made in graduations of ⅜, ½, ⅝, ¾, ⅞ and 1 inch. (See Illus 169.) Today, when inch boards are really only ¾ inch thick, the ¾-inch set of planes is the most useful to own. I buy most of my lumber roughsawn to a full inch from mills. When hand-planed, this lumber is ⅞ inch thick, so this is the size of planes that I use most often.

Both the tongue and the groove planes have a vertical fence on the left side of the stock. On "best"-grade planes, the fence's right-hand surface (the one that runs against the work) was some-

*Illus. 167. There are two types of match planes. "Best"-grade match planes have a closed tote. "Common"-grade match planes have stocks like those on moulding planes.*

*Illus. 168. The soles of a pair of match planes. The groove plane is really a fixed plow plane and has a keel like that on a plow plane. The fences of both planes are faced with thin iron strips or wear plates that increase the planes' working lifetime.*

*Illus. 169. Match planes come in graduated sizes. The pair with closed totes is ⅞ inch and was made by Greenfield (Massachusetts) Tool Co. The other three pairs are ⅜, ½, and ⅝ inch. They were all made by A. Mathieson, Glasgow, Scotland.*

times covered with a metal strip. This protected the fence from wear and accidental damage.

The groove plane (which is really a fixed plow plane) has a metal keel instead of a wooden sole. Both planes also have a stop that terminates the cutting action when the shape is complete. The stop for the groove plane is the wooden surface between the keel and the fence. The stop for the tongue plane is the bottom of the groove between the two cutting edges. In other words, the tongue plane stops cutting when the stop rides on the upper edge of the tongue it has just made.

## Selecting and Reconditioning a Plane

Examine match planes as you would moulding planes. There should be no damage to the stocks. They should not be battered or have any large checks. The irons and wedges should be the original ones.

Over the years, match planes have often been separated, and, as a result, you will see a lot more single planes than pairs. Graduated sets are even more unusual. Still, pairs of match planes are not hard to find, and usually cost about twice as much as a common moulding plane.

Even though most manufacturers graduated their match planes to the same sizes, you cannot always combine two single planes to make a working pair. The tongue thickness varies, so that not all planes are interchangeable. If you decide to assemble a pair of planes, make sure that they were both made by the same manufacturer. Of course, if you wait and find an original pair (this is not very difficult), you will not have this concern.

Check to make sure that the planes have their correct irons. The groove cutter is very heavily tapered just like a plow plane iron; however, it has a narrow tang like that on a moulding plane iron. (See Illus. 170.) If you are missing a cutter for the tongue plane, you might be able to make it from a loose dado iron. These can sometimes be found on the second-hand market. Most dealers have a "dollar box" where they sell loose parts, and one might turn up there. Otherwise, you will have to have the blank made by a blacksmith or a machinist. This is a lot of effort and unless the plane has sentimental value you would do better investing your time in finding a complete pair.

You can easily make a new wedge if this part is missing or damaged. Because the pair's irons are of different thicknesses, their wedges are not interchangeable (the groove wedge is smaller). If you have to make a new wedge, do not copy its mate.

## Sharpening the Iron

If your match planes do not have totes, remove the irons by rapping on the heel. Otherwise, grip the wedge in a vise and drive the stock off the wedge. Remember, never hammer the wedge.

Examine your groove plane's iron. Its cutting edge should be at a right angle to its sides. The lower end of the iron looks just like a plow plane's iron and is sharpened the same way. The bezel is either straight or perhaps convex like a mortise chisel's, but not concave (hollow ground) like a bench plane's.

The tongue cutter has two cutting edges separated by a gap. These two edges have to be perfectly aligned so that the shoulders on either side of the tongue are the same depth. If they are not, only the higher shoulder will make contact with the edge of the grooved board. A gap will occur on the lower side with the lower shoulder.

If you have to make a new tongue cutter from a dado iron, first anneal it. The slot in the middle can be shaped with a file. Make sure you make this slot the same size as the tongue, so that your matched boards have a snug fit. Finish heat-treating the iron in the manner discussed in Chapter 6.

Before honing the bezels, flatten both cutters on

*Illus. 170. A pair of match plane cutters. The cutter on the left makes the groove, the one on the right the tongue.*

the lapping table. Whenever you hone the tongue cutter, make sure that you apply even pressure on both sides, or over time one side may be worn lower than the other.

Return the irons to their respective planes. Be sure to align the groove in the tongue plane with the groove in its sole. The groove iron should be centered on the skate. Set the wedges and adjust the setting as you would on a moulding plane.

## Using a Plane

Match planes are designed so that when used on a board of the designated thickness, the tongue and groove are centered on the edge. When using these planes, it is best to identify one side of the boards—for example, the outside. Always place the fence on this one side. This way, the boards' surfaces will be flush when the boards are fit together. If you do not take this precaution and there is some slight misalignment between the pair of planes, the surfaces will be slightly offset. (See Illus. 171.) This is more of a problem if the boards are only finished on one side, or if the surfaces have been selected for some unique feature such as their figures.

When necessary, you can use match planes that are slightly larger than the thickness of the board. The tongue and groove will not be centered and will be shifted towards the right. (See Illus. 172.) If there is too great a shift, the groove's right edge will be thin and can easily break. (See Illus. 173.)

The right side of the tongue plane's cutter is usually slightly wider than it needs to be. So, when necessary, you can also use a plane that is slightly smaller than the thickness of the board. (See Illus. 174.) In this case, the tongue and groove will again be off center, but shifted towards the left. Once again, too much shift weakens the groove's left edge.

When using match planes that are either over-sized or undersized, do not vary more than one size, which is ⅛ inch. For example, you could use a ¾-inch plane on ⅞-inch lumber or a ⅞-inch plane on ¾-inch boards. The shifting that results makes it imperative that you always run the fence against the same surface, as the misalignment will be even greater.

Tongue-and-groove boards are matched on their edges, so the board must be mounted edge-up in your bench's side vise. Begin by placing a plane (either of the pair) on the edge of the board with the cutter just off the corner. Match planes are not sprung, so hold the plane vertically with the fence tightly against the board. As you push forward and walk the plane along the board, you should immediately see the profile take shape. The groove will cut only one narrow shaving, while the tongue cuts two which eject together out the throat. (See Illus. 175 and 176.) Repeat the process until the stop terminates the cutting action.

Usually, each board has a tongue on one side and a groove on the other. (See Illus. 177.) You can groove all the boards at one time, and set each aside as that edge is finished. Then either take them one by one to make the tongue or run one edge and immediately flip the board over to do the other side. Whichever way you choose, remember to keep track of which surface you run the fence against and to always use that surface.

Match boards are often decorated with a bead moulding that's run on the shoulder of the tongue. This simple moulding embellishes what would otherwise be a flat, uninteresting wooden surface. The bead should not be run on the grooved board. The quirk will undercut the edge, and it may become weak enough to break. If you are using an undersized set of planes and are going to mould your match boards, be sure to bias the tongue and groove towards the rear surface of the boards. This way you are sure that the shoulder is thicker than the quirk is deep.

*Illus. 171. When making matched boards, identify either the front or back side of every board and always run the fence of both the tongue and the groove planes against that side. If you do not take this precaution, your boards could end up misaligned.*

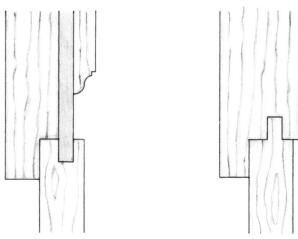

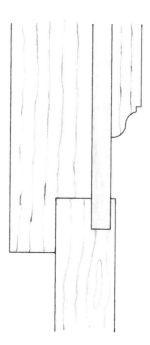

*Illus. 172. When match planes are used on boards that are slightly thinner, the tongue and the groove will be shifted to the right.*

*Illus. 173. If the board is too thin, the groove will be too far off center and the right side will be weak.*

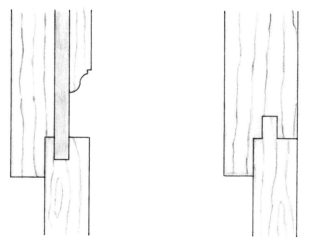

*Illus. 174 (left). If match planes are used on stock that is slightly thicker, the tongue and the groove will be shifted to the left.*

*Illus. 175. Making a groove with a match plane.*

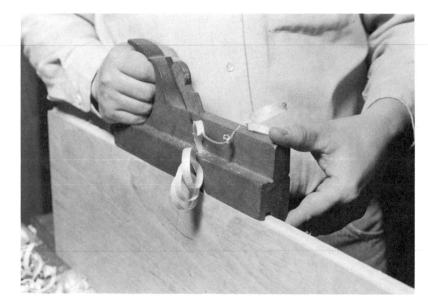

Illus. 176. Making the tongue with a match plane. Note that two chips are coming out the throat at one time.

Illus. 177. An assembled tongue-and-groove joint.

# Table-Leaf Planes

The drop-leaf table has been around for centuries. It has always been a popular type of furniture because it provides a large eating area. More importantly, the leaves drop out of the way when that large surface is no longer needed.

Woodworkers had to find a way to treat the edges of the leaves. If the edges were merely butted together, there was always a gap, especially when the leaves were down. Also, the hinge knuckles were exposed. This was not much of a problem on simple worktables, but was not acceptable on finer furniture.

There were other problems besides aesthetics. The butt joint was inherently weak, since the leaves were mainly supported by their hinges and the gate leg. If spilled food or liquids ran down into the joint, the mess would get on the back surface of the leaves where it was hard to clean.

During the first half of the 18th century, cabinetmakers developed a solution called the rule joint, which was so effective that it is still used today. It consists of two nesting profiles, a concave and a convex, that are run on the edges of the top and leaves. (See Illus. 178.) The profiles are seldom a true quarter of a circle, more usually a quarter of an ellipse. However, for this discussion I will call them the cove and quarter round. The cove is run into the bottom edge of the leaf, while the quarter round is run on the lower edge of the fixed top. (See Illus. 179.)

*Illus. 178. Table-leaf planes are really very specialized moulding planes. Their profiles are used for one very specific purpose.*

*Illus. 179. Using a table-leaf plane to make a quarter-round moulding on the edge of a tabletop.*

This quarter round is set off by a rabbet. When the edges of the leaf and top are placed together, the upper corner of the leaf fits into the rabbet on the upper corner of the top. The convex quarter round on the lower edge of the top fits into the cove in the lower corner of the leaf edge.

At no time when the leaf is hanging or is being raised does a gap appear, and the hinges are always hidden. The rule joint causes less stress on the hinges than does a butt joint, since the upper edge of the leaf rests on the projecting lower edge of the fixed top. Also, lowering the leaf gives easy access to both edges for cleaning.

Today, most woodworkers would try to make a rule joint with a router. However, it can be done just as well by hand using a pair of wooden planes called table-leaf planes. One of these planes makes the cove on the leaf, and the other the quarter round on the top. Using these planes is quick and requires no setup. Just clamp a table leaf (or the tabletop) in your bench's side vise and run the rule joint the way you would a moulding. (See Illus. 180.)

Table-leaf planes are not rare, but if you want to own a pair you will probably have to purchase them from someone who deals in specialized tools. Write to dealers and ask them to be on the lookout for a pair.

Most table-leaf planes look like moulding planes. They have the same-shaped stock and wedge. In fact, they really are two special-purpose moulding planes that are always used together and only on tabletops.

Although these tools were made in the craft-shop period by plane makers and later in the 19th century by plane factories, many also appear to have been made by woodworkers for their own uses.

Both tools have a fence. Many pairs of table-leaf planes are worked vertically. There are some exceptions, and in these pairs only the cove is sprung, not the quarter round. The cove for the table leaves has a stop. However, the quarter round for the edges of the tabletop does not always have a stop.

## Selecting and Reconditioning a Plane

Examine a pair of table-leaf planes as you would any moulding plane. Check to be sure that the

*Illus. 180. An assembled rule joint.*

two planes are a true pair. If not, there is no assurance they will nest correctly. If you are buying the tools through the mail, ask for these assurances in writing.

Since the tools are unusual, you may be willing to invest more time in restoring a pair than you would a common plane. The information needed for making replacement wedges and new cutters is found in Chapter 12.

## Sharpening the Iron

Remove the irons as you would the iron on a moulding plane. If you have to make a new cutter, the cove is very simple because the cutting edge is a quarter circle and its bezel follows this outline. The cutting edge of the quarter round is really a large ovolo with only one extra-long fillet. This fillet cuts the rabbet above the quarter round.

Return the irons to the planes. Before securing the wedge, make sure that you sight down the sole. You should see the same "halo" effect of steel projecting above the sole that you would see on a moulding plane.

## Using a Plane

Clamp the tabletop (or one of the leaves) vertically in your bench's side vise with the bottom surface facing away from the bench. The edge you are working will be pointing upwards. Remember that the quarter round is made on both edges of the top, while the cove is run only on the mating edge of each leaf. So, you will mould the top twice and the leaves once. When you turn the top over to mould the other side, make sure that the bottom surface continues to face away from the bench. Ignoring this could cause a disaster.

If your table-leaf plane is not sprung, hold the stock vertical with your right hand. Place the fence against the board's bottom surface and hold it tight with your left hand. Place the cutting edge just off the corner of the board and push forward.

With each pass you will see the profile develop. The quarter round is the easiest profile to make. While it is being formed, the iron is also cutting the rabbet. This helps the tool to track properly and keeps it from wandering. However, the quarter round does not usually have a stop. Cease planing when the profile is fully formed along the entire edge.

You are more likely to have trouble with the cove. Watch each pass and make sure that the fence stays tight against the work. If it wanders, decrease the downward pressure on the tool so that the next few passes do not deepen the cut, and increase the pressure against the board so that the cove widens back to its full width. When the fence is once again running against the work, finish the cove. This is less of a problem if your cove plane is sprung. You automatically push the plane against the wood at a 45-degree angle, so it is less likely to wander.

# Panel-Raising Planes

As already discussed, wood experiences seasonal movement; it expands and contracts as it balances its own moisture content with the relative humidity of the surrounding air. As a result, you will encounter some problems when trying to make a wide wooden surface such as a wall or a door.

We already know how matched boards accommodate seasonal expansion and contraction.

There are other solutions, such as raised panel doors and walls, and feather-edged sheathing. Although these architectural features were developed several centuries ago, they have proven to be successful and remain very popular today. They are appreciated by customers because of the way they look, and they are appreciated by woodworkers because they are very practical ways to use wood.

The raised panel is aesthetically successful because of the way it uses light and shadow. Even under the full illumination of direct bright light, the raised field is set off by its bevelled perimeter. Under oblique light, the bevel is in shadow; this "articulates" the surface and gives the appearance of great depth.

A raised panel is placed in a wooden frame of vertical stiles and horizontal rails which have grooved edges. The panel's four bevelled sides are drawn out to a feather edge that fits into these grooves. This allows the panel to float unconstrained in a frame of stiles and rails. Because the panel's expansion and contraction is accommodated by the grooves, this is a very practical way to produce a wide wooden surface that does not develop gaps when the wood contracts.

Stile, rail, and raised-panel construction could be used for entire walls, over mantels, and for wainscoting. It also makes a very attractive, but lightweight, door. (See Illus. 181 and 182.)

Feather-edge sheathing is used to make interior walls of vertical or horizontal boards. The feathered edge is the same profile as that found on raised panels. The thin edge is fitted into a groove run into the abutting edge of the next board. (See Illus. 183.) Different effects can be obtained by running a feather edge and a groove on each board or a feather edge on both sides of every other board and grooves on both edges of the alternating boards. Feather-edged boards can also be used to make interior doors. The boards are held together with battens. (See Illus. 184.)

Today, raised panels are usually made on a table saw or a shaper. However, both methods are clumsy, require too much setup, and need too much cleanup. I prefer to use a panel-raising plane. This tool is quick and easy to use, and makes a perfect raised panel.

A panel raiser is really a special moulding plane. The profile it cuts is the shape of a panel's feathered edge, and so its sole is shaped like that on a moulding plane. The stocks of most panel raisers are similar to that on a jack plane and usually have an open tote. (See Illus. 185.) Some English panel raisers have a stock like that on a large smooth plane.

Panel raisers have the same type of throat as bench planes, and so the chip is ejected out of the top rather than to the side. The throat has all the features described in Chapter 9: the bed, wear, cheeks, and abutments.

Each time you raise a rectangular panel, you have to do two sides that are end grain. So, like the iron on a fillister, the iron on a panel-raising plane is set askew. As a result, the wear, the bed, and the abutments are not at a right angle to the sides of the stock. The wedge, which looks like those used on bench planes, also has bevelled edges, so its cross section is a parallelogram. (See Illus. 186.) The top edge of the wedge and the ends of its two legs form another parallelogram, so, as a result, making and fitting a replacement is a challenging project.

Panel raisers have both a fence and stop. These

*Illus. 181. A raised panel door. (Photo courtesy of Wentworth-Gardner House, Portsmouth, New Hampshire.)*

*Illus. 182. A raised panel wall and raised panel doors in the upstairs hall of the Wentworth-Gardner House. This elaborate work clearly illustrates why this house, now a museum, is considered one of America's finest Georgian mansions. Combined with mouldings, raised panels create an elaborate interplay of light and shadow.*

*Illus. 183. A wall made of featheredge sheathing separates rooms in the domestic servants' quarters of the Wentworth-Gardner house. The featheredge sheathing was made with a panel-raising plane. The pieces of paper on the wall are laundry lists dating from the 1860s.*

*Illus. 184. A pair of featheredge doors in the servants' quarters at the Wentworth-Gardner house. The closed door (left) shows the featheredged boards. The open door reveals the two horizontal battens that hold the boards together.*

*Illus. 185. A group of panel raisers. Note that their irons are set askew, so that they can cut across grain more cleanly.*

*Illus. 186. A panel raiser's wedge has bevelled edges, so its cross section is a parallelogram. Making a new one that fits correctly is a challenging woodworking project.*

planes are usually sprung so, like a sprung moulding plane, the angle of the fence tells you the correct position in which to hold it. Spring lines are often scribed into the toe.

It has been suggested by some authors that panel raisers should come in sets, one left- and the other right-handed. I have never seen a right-handed plane. All panel raisers are meant to be fenced on the left-hand edge of the panel, and work so well that a pair of planes is not necessary.

Most panel raisers make just one shape and one size of a feathered edge. The profile consists a small bevelled drop from the raised field and a

long, straight bevel (called the raise) that extends out to the narrow edge. (See Illus. 187.) Others make a raise that is slightly concave. Some panel raisers actually make a small moulding profile (a bead or a small ogee) along the field's dropped edges. This is sometimes called mould-on-raise, and during the Georgian period was used in very fancy mansions on doors that were sometimes made of mahogany. The planes that made this mould-on-raise are still occasionally found, but are not common.

Some panel raisers are adjustable and have a fence that moves back and forth to increase or

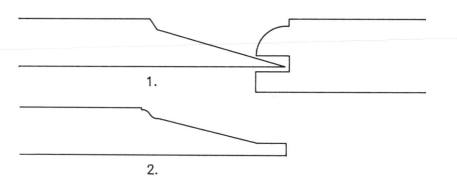

*Illus. 187. 1 shows the cross section of a raised panel. The edge of the panel is held by a groove running into vertical stiles and horizontal rails, and should fit snugly. To ensure clarity, the groove in the illustration has been drawn oversized. 2 shows another variation of the featheredge. It has also been drawn as moulding-on-raise.*

decrease the width of the raise. This fence is similar to the type found on fillisters, and is adjusted by loosening two screws. This adjustable fence allows you to make a more narrow raise on a small panel, so that the field and the raise remain in proportion to each other.

Pine and other soft woods were most commonly used to make interior architectural features, so most panel raisers have a common pitch. Hard woods like poplar, mahogany, and even oak were sometimes used to make some doors, so some panel raisers have a York pitch (50 degrees). Like most other moulding and special-purpose planes, panel raisers do not have a capped iron.

Since raised panels fell out of popularity in America before the Industrial Revolution, they were not made in large numbers, if at all, by plane factories. Most American panel raisers were hand made by plane makers working in the craft-shop tradition, and all are what are referred to as 18th-century planes.

In England, panel raisers were used much later, and some English plane factories made these tools. So, you will find handmade and factory-made English panel-raising planes.

## Selecting and Reconditioning a Plane

Panel raisers are not rare, but neither are they common. They are generally expensive, but if you make more than an occasional raised panel you should buy one of these planes. Panel raisers that were made by known plane makers are highly prized by collectors. In fact, some signed panel raisers are expensive enough so that those who make fakes may soon start to duplicate them. You are assured of a genuine panel raiser if you buy a less expensive, unmarked one, as it is not worth a faker's time to make one of these. On the other hand, panel raisers have recently been reproduced by legitimate tool manufacturers, though it will probably be years before these copies enter the second-hand tool market.

Before buying a wooden panel raiser, examine it as you would both a bench plane and a moulding plane. If you are buying the tool through the mail, make sure you can return it if it turns out to be unsuitable for use.

The tool should not be battered or have any major damage. Sight along the sole to see if the stock is warped or wracked. It is very difficult to correct this problem on a moulding plane because the sole is so narrow and its details so small. However, the sole of a panel raiser is much wider and made up of just three angled surfaces: the fence, the raise, and the drop. I have successfully removed a small amount of distortion with a metal bench rabbet plane and a metal shoulder plane. Both these tools are described in the next chapter. However, you should not make any changes or corrections to the sole until you are very well acquainted with wooden planes. Otherwise, you may ruin the job and an expensive tool.

## Sharpening the Iron

A panel raiser's cutter is a single iron (no cap) that's shaped like a bench plane iron. If you find a panel raiser that is missing its cutter, you can sometimes find single irons or have one made for you. You can even make a very serviceable replacement from a spare bench plane iron, although it will have a slot cut in it. This will not affect the way the tool works, but it will affect its resale value. The market places the highest value on tools that are completely original, and the replaced iron will be very obvious.

The bezel on a panel-raiser iron has two

straight cutting edges, and each has its own relief. (See Illus. 188.) The smaller of these edges cuts the drop, while the other makes the raise. If you are making a replacement iron, spray layout fluid on the front surface and trace the outline of the sole with a sharp stylus. Grind your iron to this shape and then make the bezels. Remember that each cutting edge has to have its own relief. So, the intersection of the finished bezels will usually form a short line that is nearly parallel to the sides of the iron.

*Illus. 188. A panel raiser's cutter has two straight cutting edges, each with its own relief.*

Always flatten the surface opposite the bezel, removing any pitting or other blemishes. Next, hone the bezels and, if necessary, strop lightly.

Place the iron back into the throat and set the wedge with enough pressure to hold the iron. Sight down the sole. When the iron is properly adjusted, you will see the cutting edge running parallel to the surfaces that make the raise and the drop.

Making a replacement wedge for a panel raiser is more difficult than for a bench plane because it requires more complicated geometry. This is because besides tapering from its top end to the points of its legs, the cross section of a panel raiser's wedge is also a parallelogram. So, first make a blank with correctly bevelled sides. Next, taper the end that will fit in the mouth. You may throw away several blanks before you fit one that creates even pressure under the abutments. When the wedge fits satisfactorily, you can cut out the legs and add the finishing details.

# Using a Plane

Carefully select a piece of wood for a panel. Panel raisers were not intended to work very hard wood or highly figured wood. Use a select grade with grain that runs in one direction so that both sides can be worked without tearing.

Clamp the wood you will make into a panel (or feather-edged sheathing) onto the workbench top, with the left-hand edge hanging slightly over the edge of the bench. Use the panel raiser like a moulding plane. Place the fence on the left edge of the board, with the mouth just off the near corner.

Look at the fence. It will tell you at what angle to hold the plane. If it does not have spring lines, make your own. The plane makes a wide cut and requires a decisive push. You will have to walk alongside feather-edge boards and long panels, but can reach across the short length of a panel. The end-grain shavings will crumble. The edge-grain shavings, however, are not broken by a chip breaker and will shoot up out of the mouth.

When using a panel raiser, begin on one end of the panel. This is the grain end. Clamp a piece of scrap between the corner of the panel and the bench dog. The scrap will prevent the far corner from chipping as the plane cuts across it, and should be the same thickness as the wood.

Raise that edge. Next, raise the other length of end grain, once again using a piece of scrap (usually the same piece) to prevent chipping of the far corner. When the two ends are done, raise the two sides.

The intersections are critical. They must meet so that the line formed by them runs straight from the corner of the field to the corner of the panel. If that line is biased to one edge, rather than meeting at the corner, take another stroke or two with the panel raiser on the edge towards which the line of intersection is biased. (See Illus. 189.) This will correct it. You will discover that it is easier to make these details accurately with a plane than with a table saw or shaper.

Another type of panel was popular during the second quarter of the 19th century, during the architectural period that is known as the Greek Revival. This type of panel is still seen today. It has a raised field, but the edges are not bevelled. Instead, a rabbet is run on all four sides. This is done with a moving fillister rather than with a panel raiser.

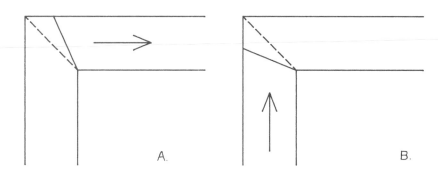

*Illus. 189. When raising a panel, make sure that the finished corners form perfect mitres, as represented by the dotted lines. If the two adjacent bevels are shifted as shown in A, remove more material by using the plane in the direction of the arrow. If the junction is shifted as shown in B, correct that shift by planing in the direction of that arrow.*

A.

B.

# Compass Planes

The very word plane suggests that a tool is used to work areas that are straight and flat. Most planes are usually run on straight edges or flat surfaces. However, much furniture and some architecture contains parts that are curved. To smooth these parts, cabinetmakers and joiners rely on a plane with a curved sole called a compass plane. This term is a corruption of the earlier name for this tool, "encompassing plane."

Compass moulding planes and a variety of special-purpose compass planes were made for such crafts as carriage-making. However, here we are interested in the compass plane used by cabinetmakers for making furniture and those joiners used for making architectural elements. Most of these wooden compass planes look very much like a coffin-shaped wooden smooth plane. The difference is the convex-curved sole that allows you to run it down a concave surface. (See Illus. 191.)

Wooden compass planes come in many sizes that range from those as large as a smooth plane to some that you can cradle in the palm of your hand. They vary in both the size of their stock and the radii of their curved soles.

Not all compass planes were used for smoothing. Sometimes they were intended for excavating. In this case, the tool's sole is also rounded from side to side (laterally) as well as front to back (longitudinally). I use an ebony compass plane whose sole is contoured in both directions to finish shaping Windsor chair seats. I know of no wooden compass planes that are in-curved—that is, that have a sole that is concave front to back.

Since a compass plane has a body that is similar to that of a smooth plane, it is not surpris-

*Illus. 190. Using a panel raiser.*

*Illus. 191. Three compass planes. The largest one is only curved front to back. The other two are curved side to side and front to back. This indicates that they were intended for excavating. Since both of their cutters were made from old files, the planes were probably made by a woodworker, rather than a plane maker.*

ing to find that they share many other features. The throat is the same, and the parts have the same names. For example, on both compass and smooth planes a two-legged wedge slides under the abutments to hold the iron secure against the bed. Some compass planes are so small that the wedge's legs are slightly larger than toothpicks. In this case, the space between the legs is not removed, but only relieved so that the wood between them forms another incline steeper than that of the wedge. (See Illus. 192.) This way, the wedge does not interfere with the passage of the chip, and does not have fragile legs.

## Selecting a Plane

Before purchasing a wooden compass plane, make sure that it is not a converted smooth plane with a sole that has been reshaped to a radius. If it is, wood has to be removed from the toe and heel area of the sole. This will open the mouth. If you do find a converted smooth plane that appeals to you (perhaps because it is in excellent condition),

*Illus. 192. The wedges of some minute compass planes are so small they sometimes have very fragile legs or do not have legs at all. One of the legs on the wedge on the right has, as a matter of fact, broken. A steeper incline is sometimes cut into the upper surface of the wedge to help guide the chip.*

close the mouth. Make an inlet in the sole and fit a patch as described in Chapter 9.

I have observed no specific problems that are unique to wooden compass planes. When they have problems, they tend to be of the general variety, for example, excess wear and broken or missing parts. Examine any plane you are considering buying for these defects.

## Sharpening the Iron

Sharpen a compass plane blade as you would that of a smooth plane. If the sole is curved laterally (across the sole) as well as along it, the iron will be sharpened like that of a wooden jack plane. The cutting edge will be ground to a radius.

If you have to grind the cutter (or make a new one) for an excavating compass plane, spray the forward surface with layout fluid and trace an arc that has a radius that's slightly smaller than that of the sole.

Return the sharpened iron to the plane and set it as you would a smoothing plane iron.

## Using a Plane

A compass plane will work in a curve with a radius that's longer than that of its own sole. However, it will not work on a curve with a radius that's shorter than the radius of its own sole.

Because compass planes are used on concave wooden surfaces, you have to be aware of the wood's grain. As you work down the curve, you will be cutting at a slight angle across the end grain. However, once you reach the bottom of the curve and start up the other side, you will discover that the compass plane is chipping or tearing. (See Illus. 193.) This is because you are now running into the grain. To prevent this, run the compass plane down only as far as the bottom on the curve, lifting the tool slightly at the end of

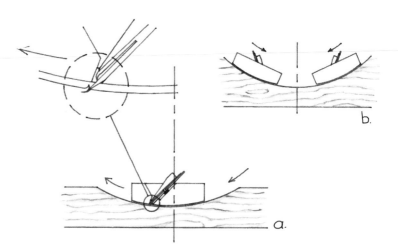

Illus. 193. When using a compass plane to smooth a concave curve, plane as shown in b: down each side, ending the stroke at the bottom of the curve. Otherwise, the cutting edge will tear as it runs into the grain on the opposite side, as shown in a.

each pass to cut the chip free of the wood. Work the other side the same way.

Because you are working end-grain, a compass plane is inclined to chatter. You can prevent this by tightening the muscles in your forearms so that your wrists are as rigid as possible.

# Nosing Planes

The nosing plane is also commonly mistaken for a moulding plane. The confusion results because these two types of tools look very much alike. Nosing planes have the same-shaped stock, and their concave soles lead many dealers to label them as hollows or beads. (See Illus. 194.)

Illus. 194. Three nosing planes. The plane on the right is ½ inch and is unmarked. The middle one is ⅞ inch and was made by Auburn Tool Co., Auburn, New York. The plane on the left is 1¼ inches and is also unmarked.

The sole of a nosing plane is a full half round like that on a bead plane, but like a hollow plane it does not have a fence or stop. These planes are not sprung.

Nosing planes are used for rounding the edges of such architectural features as stair treads and windowsills. They were made in graduations that correspond to standard thicknesses of lumber, for example, ½, ¾, and ⅞ inch. You will usually find the size of the plane stamped into the heel. Today, a 1-inch board is really ¾ inch, and this is probably the size nosing plane you will most commonly use. However, when wooden planes were being made, a 1-inch board was thicknessed to ⅞, and this is one of the more common sizes you will find on the second-hand market.

## Selecting a Plane

Examine a nosing plane as you would a moulding plane. After looking at both the stock and the wedge, check the sole to be sure it is not warped or wracked.

Make sure that the plane has not been altered and is not an alteration itself that has been made from a moulding plane. Look at the throat to see if it has been expanded. As will be explained shortly, a nosing plane cuts two separate shavings, and they may choke in an altered throat.

## Sharpening the Iron

Do not drive the wedge when removing the iron. If you cannot pull the wedge loose with your hand, rap the heel of the stock with a mallet. If this will not work, grip the wedge in a vise or drive the iron out of the mouth by hammering the tang.

The iron of a nosing plane is sharpened like that of a hollow or a bead plane. If you measure the distance across the plane's concave profile, you will find that it is just slightly wider than the board on which the tool is meant to be used. I will explain why later.

The first step in sharpening the iron is to flatten the side opposite the bezel. Next, hone the bezel with a slipstone. A round one, such as the ceramic file that I use, is perfect for this job.

It is not necessary to hone the square ends of the cutter on either side of the concave cutting edge. Although they are each relieved by a bezel, they do not engage the wood.

If you have to make a replacement iron or restore the profile on a worn cutter, spray the forward surface with layout fluid. After the spray has dried, return the iron to the tool. Trace the shape of the sole in the dried fluid and reshape to this line. Illus. 195 shows how the bezel of a nosing plane cutter is shaped.

# Using a Plane

As with a moulding plane, when the cutter of a nosing plane is set correctly you should see a "halo" of the cutting edge projecting above the sole.

Use the plane on the edge of a board. Surface the board you are going to work to the same thickness as the size of the plane and joint the edge. Clamp the piece of wood in the front vise of your bench and set the plane on the edge. The tool is not sprung, so hold the stock vertically, in line with the board. The mouth should be set just off the corner of the board, ready to engage the wood with the first pass.

The sole is just slightly wider than the thickness of the wood you are working. On my ½-inch nosing plane, the distance from corner to corner is ⁹⁄₁₆ inch. On my ⅞ inch plane, it is a full inch. This means that the plane will straddle the edge of the board. This is a necessary feature because there is no fence to keep the tool on the work.

During your first pass, the iron will cut on both sides and should remove the square corners from the board's edge. The next pass will also cut two shavings and give the corners a slight rounding.

Each run of the plane will cut two shavings that gradually widen until the profile is complete. (See Illus. 196.) The blade then cuts a single, wide chip. The sole has no stop, so a single shaving, cut from the entire round edge, is your signal that the profile is finished.

*Illus. 195. The bezel of a nosing plane cutter is not an even width around the entire curve. There is more relief at the top of the curve than along the sides. The two lower ends are relieved, even though they do not engage the wood. They do not have to be honed.*

*Illus. 196. Using a nosing plane. The plane first cuts two shavings, one from each corner of the edge, that gradually increase in width until they merge into one wide shaving. At this point, the profile is complete.*

On most applications for the nosing plane, such as stair treads and windowsills, the profile also has to be run on both ends of the board. The ends are, of course, end grain. If the plane is properly sharpened, it will cut them as well.

When planing across end grain, you do have to worry about splintering the far corner. Clamp a piece of scrap wood the same thickness as the board against the far edge. This scrap will splinter instead of the piece you are working.

# Forkstaffs

The compass plane is used for working concave surfaces. Another special-purpose plane, the forkstaff is used for round work. Like compass planes, many forkstaffs (or forkstaves) have stocks that are the same shape as those for smooth planes. Once again, the difference is in the shape of the sole. It is concave laterally (from side to side).

Shipwrights referred to the forkstaff as a spar plane. When you consider that a spar is a stout, round wooden member used to support the rigging of a ship, you'll realize that the term spar plane explains the tool's purpose: to do round work that is either too big or too long to be done in a lathe. For example, I use my forkstaffs for rounding Windsor chair bows. These planes are used to do heavy work, and so many of them have a double iron. (See Illus. 197.)

Forkstaffs are a common item on the second-hand market. If you want to use a forkstaff, you shouldn't have much trouble finding one, and it should not be expensive.

## Selecting a Plane

Examine a forkstaff as you would a smooth plane. Make sure that the stock is sound and that the throat, cheeks, and abutments are not damaged.

Many of the forkstaffs you will come across started out as smooth planes and were later adapted. Hollowing the sole can open the mouth so wide that these altered tools are inclined to tear the wood they are working. An original forkstaff will have a mouth that is nearly the same width along the entire length of its curve. Don't buy a forkstaff that was made from a smooth plane. In contrast to a compass plane (that is an altered

*Illus. 197. Both these forkstaves are made of live oak and are unmarked.*

smooth plane), closing the mouth of a forkstaff with a patch is not an easy job.

Before buying a forkstaff, also check the sole. If it is badly worn from use, you cannot joint it and start over again.

## Sharpening the Iron

Remove the iron from a forkstaff by rapping the heel with a mallet. The shock will loosen the wedge. Begin the sharpening process by flattening the side opposite the bezel. Hone the cutting edge with the round edge of a slipstone. I use a round ceramic file for this purpose.

If you have to make a new iron (or reshape one that is damaged or worn), follow the same technique as used for a nosing plane. Spray the surface opposite the bezel with layout fluid and let it dry. Place the iron back in the plane. Trace the outline of the sole in the dried fluid and reshape to this mark. Next, grind the bezel. Unlike the nosing plane, the bezel of a forkstaff iron is parallel to the cutting edge across the complete width of the cutter, producing sharp, pointed corners. (See Illus. 198.)

If the tool has a chip breaker, be sure to periodically hone its front edge. This will ensure that it lays tightly against the cutter and prevents the chip from getting stuck under it.

## Using a Plane

You can round a piece of wood with a forkstaff as long as the wood's cross section has a radius that is shorter than the radius of the plane's sole. In other words, the wood has to have a smaller curve than the curve of the plane. If you try to use the

*Illus. 198. The bezel of a forkstaff cutter is concave like that of a nosing plane, but is the same width across its entire length.*

forkstaff on a piece with a cross section that has a longer radius, the sharp corners of the cutter will make contact rather than the cutting edge. This will mar the surface you are trying to round.

A forkstaff is generally used on long pieces of wood that are at least nearly square in cross section. If you have a lot of wood to remove, rough-round it with a drawknife (see Chapter 16). At least use a drawknife to knock off the corners. Clamp the wood in a vise or on a bench top in a manner that will allow you to run the forkstaff along its entire length. If you have to stop in the middle of each stroke, it will be more difficult to make the piece a continuous round one.

Be sure not to run the plane along the same path more than once or twice. Remember, you are rounding and have to move the plane steadily around the piece's entire surface. Working the same surface for too long will result in an oval cross section. It is a good idea to look at both ends of the wood to see how its cross section is developing.

After you have made several passes, each time moving the plane around the wood's diameter, you will eventually come into contact with the bench top. Thus, it is necessary to regularly loosen the wood and rotate it to keep the surface you are working directed upward. (See Illus. 199.)

*Illus. 199. Using a forkstaff to round a piece of wood. Be sure to rotate the wood frequently to avoid making it oval.*

# Mitre Planes

As was explained in Chapter 9, the blade's pitch (or bedding angle) varies with the plane's purpose. The irons of most woodworking planes are bedded at 45 degrees. Planes that work harder or more figured woods have one of the higher pitches.

On the other hand, some planes have a pitch that is lower than 45 degrees. A low bedding angle is a good indication that the tool is meant to be used on end grain where the cleanest surface is obtained by a low-angled paring cut. You may know from experience that if you use a bench plane on end grain it often creates dust rather than a shaving.

The wooden mitre plane has a low bedding angle, usually about 35 degrees. In fact, its name implies one of its primary functions: cleaning mitred joints so that they fit neatly. This is not heavy work, so they do not have a chip breaker and usually have a single tapered iron.

You will come across mitre planes with two types of bodies. The first type is coffin-shaped like a smooth plane. Others have straight sides that are square with the sole. This shape allowed these planes to be used on mitre-shooting boards. (See Illus. 201.)

You may have to joint a mitre plane's sole. Follow the instructions in Chapter 9. If you want to use your mitre plane on a mitre shooting board, make sure that the finished sole is still square with the sides. Otherwise, your mitres will not fit correctly.

Since a mitre plane is used for fine trimming, it has to have a tight mouth. If necessary, close it with a patch as explained in Chapter 9.

## Selecting a Plane

Wooden mitre planes are not as common as bench planes, but neither are they hard to find. Most dealers confuse them with smooth planes and, as a result, they are not expensive.

Any mitre plane you buy should be in good condition. You should not have to do any more than clean the tool, joint its sole, and sharpen the cutter. Examine it for flaws the same way you would examine a smooth plane, compass plane, or forkstaff.

## Sharpening the Iron

Mitre planes have a single iron that is used with the bezel down, in the manner of a smooth plane. Because the bedding angle is so low, you will have to create more relief when grinding the bezel. (See Illus. 202.) Otherwise, the bezel's heel will rub on the surface of the wood and keep the cutting edge from making contact. The bezel on my wooden mitre plane is about 26 degrees. Such a steep angle means that the edge is more fragile and can be damaged by heavy use. However, a mitre plane is used for delicate work (shaving end grain), usually as part of the final fitting of joints such as mitres. The edge is strong enough for this type of work.

## Using a Plane

A mitre plane is used for trimming, so it must be able to pare very thin shavings. Adjust the tool

*Illus. 200. This mitre plane has a steel outer shell that is filled with rosewood. It was made by Rt. (Robert) Towell, London, England and is bedded at about 20 degrees. Bench planes, rabbet planes, and shoulder planes were made this way. Wood-filled planes are extremely precise and are the best planes ever made. While plentiful, they are quite expensive and have to be bought from tool dealers. (Plane in photo provided by Malcolm MacGregor.)*

*Illus. 201. These three mitre planes all have irons that are bedded at 35 degrees. The plane on the right has a coffin shape. The other two have straight sides that allow them to be used on a mitre shooting board.*

*Illus. 202. Because a mitre plane's cutter is bedded at a low angle, the bezel has to be longer than that on a bench plane. If not, it might not have sufficient relief, and the heel would make contact with the wood rather than the cutting edge.*

the same way you would a smooth plane. Advance the iron by tapping the upper end, and retract it by tapping the heel. To make a lateral adjustment, tap the side of the iron to move it in the required direction.

Besides the mitre shooting board (Illus. 203), the mitre plane is often used in conjunction with another tool called a mitre jack. (See Illus. 204.) Although they are not generally known in the United States, mitre jacks are still made and used in Europe. This device is a special wooden clamp that is made up of two jaws, each with a face that is cut at precisely 45 degrees. These jacks are not a common second-hand tool. On the other hand, they are not rare, and you will not have trouble finding one if you contact specialized tool dealers.

# Sash Planes

Like all other woodworking jobs, window sash can be done by hand. I know woodworkers who

custom-make sash this way for historic restorations. However, except for these special situations, sash is made in mechanized and automated factories. But woodworkers can use glass panes in places other than windows, like, for example, in bookcase doors, secretaries, and cupboards. The special-purpose plane that makes sash is a very handy tool for doing these jobs. A sash plane makes the moulding and rabbet on the muntins, as well as on the stiles and rails of the frame.

Sash makers use several other tools to help them quickly cut the large number of precise joints in sash; for example, they will use a matching sash coping plane and a mitre template. These tools originally came in sets with the sash plane. The sash coping plane was used to undercut tenon shoulders to the reverse of the muntin's moulding profile. The coped shoulder would nest over the moulding, resulting in a perfect joint. The matching mitre guide (or template) was used to make mitred bridle joints where muntins intersect. Its underside is moulded to the reverse of the muntin profile, which allows it to be placed on the muntin and slid into position. The four corners are angled at 45 degrees. Each end forms a vertical V, like the bow of a boat. These mitred corners are used to guide the saw when it's being used to cut the joints.

Although I have had to make and repair an occasional piece of sash, I do not own a template and coping plane and cannot write about them with any authority. If you want to own a sash plane with a matching coping plane and template, contact a specialized tool dealer. These tools are not rare, and can be found on the second-hand market. Try to buy the sash plane, matching coping plane, and template together as a set. A sash plane by one maker might not match perfectly the coping plane or the template made by another maker.

The most basic piece of sash is one sheet of glass secured in a frame. The frame has vertical stiles and horizontal rails. Some sash has two panes divided by a single vertical muntin, while others have four panes separated by crossed muntins. These one-, two-, and four-paned windows are the types of windows that have been produced over the last one hundred years.

Before the Industrial Revolution (when it was more difficult and time consuming to make large

*Illus. 203. Using a straight-sided mitre plane on a mitre shooting board.*

*Illus. 204. Using a mitre plane and mitre jack to trim a mitre. Place paper over the mitre jack to protect it from the plane. Here the paper was removed to better show the jack.*

sheets of glass), sash had smaller panes and more of them. In these windows, the panes are set into a grid of muntins reminiscent of a tic-tac-toe square. Six-, nine- and twelve-paned windows were common. These types of multipaned windows are still made and are often used in colonial-style houses. (See Illus. 205.)

The four pieces of sash frame (the two vertical stiles and the two horizontal rails) have a moulding profile run on the inside edge of the surface that faces into the room. The surface that faces outward has a rabbet run into its inside edge. The narrow muntins have the same moulding profile run on both edges of the side that faces into the

room. Outward, a rabbet is run on both sides. The rabbets on the back of the muntins form a thin spline that separates the panes of glass. (See Illus. 206.)

The rabbets (on the muntins and adjacent stile and rail) form an individual frame that is lined with a bed of putty. The pane is set into this frame and held in place with glazier's points and a bevelled bead of putty that is both airtight and watertight.

The mouldings on the stiles, rails, and muntins on the inside surface of the sash create interesting patterns of light and shadow. During the centuries when sash was made by hand with sash

*Illus. 205. A "nine over six" window in a Federal period house in Portsmouth, New Hampshire.*

*Illus. 206 (below). A cross section of the multipaneled window shown in Illus. 205. The stiles have the same moulding profiles as the muntins (ovolos). The rear edges of the muntins and stiles are rabbeted to hold the glass lights. A beveled bead of putty holds the glass in place and creates a weatherseal. The sash plane is run on one edge of the stile and on both edges of the muntin.*

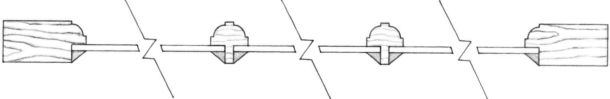

planes, different muntin profiles (like mouldings) corresponded to different styles of architecture. There are typical Georgian, Federal, Greek Revival, Gothic, and other Victorian muntin profiles.

You can buy sash planes on the second-hand tool market that make any one of these profiles. Planes that make Georgian and Federal shapes are usually products of the craft-shop tradition of plane making. Like moulding planes, they are referred to as "18th century" planes, and they too,

display their maker's individual characteristics. By the time Greek Revival and Victorian profiles had become popular, sash planes were being made in factories. Like moulding planes, factory-made sash planes are all standardized.

No matter when a muntin profile first became fashionable, most of them are still used and recognized today. Unless you are reproducing period woodwork or doing restoration work, do not be concerned with the profile you use. Any sash plane you own will make a shape that will be suitable for your project, no matter how modern its design.

There are two types of sash planes: one that makes a profile of a fixed width, and another that makes profiles of various widths. (See Illus. 207.) This second type is also called a sash fillister. Fillisters are actually two planes—a moulding plane and a small skew rabbet plane—held together in one of several ways. They can be held together by threaded arms and wooden lock nuts, square arms and wedges, or with an iron bolt and captive nut.

Some fixed-sash planes have a single iron, but most (as well as all fillisters) have two. The iron on the left side of the plane (near-side when in use) cuts the profile. The iron on the right (bench-side) cuts the rabbet. (See Illus. 208.)

The two irons make two cuts, one on each side of the plane. There are also two throats, one on each side. Thus, the chip from the rabbet is ejected onto the bench, and that from the moulding onto the floor. If there is a single cutter, two cuts are still made, but there is just one throat on the right (far side), and both chips are ejected onto the bench.

Because a sash plane is essentially a moulding plane joined to a skew rabbet, its sole has some predictable features. (See Illus. 209.) Looking at the left-hand-side of the sole and moving right, you will note the following: the left side of the sole has a fence that makes the plane track properly. Next comes the moulding profile and its cutter. Between the irons is a space that does no cutting. This is the plane's stop. Finally, on the right side is a flat sole (like that found on a rabbet plane) about ½ inch wide. This cutter is set askew (only if the plane has two irons), just like the cutter on a rabbet plane. The skewed edge helps minimize tear-out.

When sash was made by hand with sash planes, the craft was a specialized activity. Sash was not made on site, but in millwork shops. Usually, when a particular type of tool was used by only a small number of specialized craftsmen, that tool is rare and difficult to find on the second-hand market. However, sash planes are plentiful. This means that many woodworkers owned them. They must have been used for making an occasional piece of sash such as a skylight, a cellar window, or a built-in china cupboard. I own a sash plane and have used it for these particular purposes.

# Selecting a Plane

Before buying a sash plane, examine it as you would both a moulding and a skew rabbet plane. Make sure that all the pieces are the original ones, that the body is not wracked, that the fence is not worn and that neither throat has been altered.

If you are buying a fillister, make sure that the movable parts are all the original ones and are undamaged. Try adjusting the plane, or even take the two sides apart. If the two halves are secured with wedges, loosen them by tapping on the end with the round lobe.

If the plane has a single iron, examine this cutter very carefully. If it is badly worn and you have to rework its cutting edge, this will prove to be twice as difficult as reshaping a moulding plane iron.

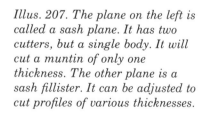

*Illus. 207. The plane on the left is called a sash plane. It has two cutters, but a single body. It will cut a muntin of only one thickness. The other plane is a sash fillister. It can be adjusted to cut profiles of various thicknesses.*

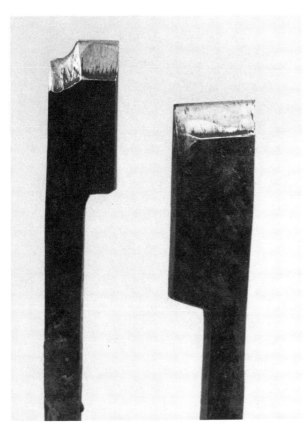

*Illus. 208. The cutters from a sash fillister look like the cutters from a moulding plane and a skew rabbet plane. The moulding profile of the cutter shown here is an ovolo, a shape that was popular during the Federal period (1790–1810).*

*Illus. 209. A sash fillister is really a combination of moulding plane and skew rabbet plane. The fence is on the left, and the stop is the surface between the cutters.*

# Sharpening the Iron

Since the sash plane makes two cuts at once, both cutters (or both sides of a single cutter) have to be equally sharp. Take the irons out of the plane by rapping the heel with a mallet. This should loosen both of them. You cannot loosen the wedges by gripping them in a vise, since there are two of them and less than ½ inch of space in between. However, if you can loosen one wedge, you can free the remaining one this way.

If there are two cutters, sharpen them as you would a moulding plane and a skew rabbet iron. Begin by flattening them. Hone the moulding cutter with slipstones that are shaped so that they will reach into every detail, concave and fillet. I use ceramic files for this purpose.

Hone the rabbet cutter on a flat stone. Remember, the bezel is at an angle. Distribute the pressure evenly across its width so as to not wear one side more than the other.

If either cutter is badly worn so that it needs to be reshaped, spray the forward surface with layout fluid. Return it to the plane and with a long scratch awl or other fine stylus trace the profile from the shape of the sole. Reshape to this line.

If your plane has a single cutter that is badly worn, you will have to reshape both sides. If you try to do just one side, the other side will be set so deeply that it will choke. Spray the entire forward surface with layout fluid. Return the iron to the plane and set it heavier than you would for use. This will expose enough surface to trace the new profile. Reshape to this line.

On a single iron, pay special attention to the gap between the two cutters. The gap corresponds to the plane's stop. Make sure that the iron does not project below the sole at this point. If it does, it will stop the cutting action before the profile is complete. By dragging on the wood, it may ruin the work that has been done. If you need to increase the depth of this gap, use the edge of a mill file.

# Using a Plane

Return the irons to your plane. Each side will have to be set individually. Do this just as you would for both a moulding and a skew rabbet plane. As you sight down the sole, you should see a "halo" of the cutting edge rising just slightly above the sole on the moulding side. The cutter on the rabbet side should be set to the same depth, and its left side (the one nearest you when in use) should be even with the inside edge of its sole.

The quality of the wood you use is critical. It must be select-grade pine (or an equivalent species) with a straight grain that has only one direction. You do not want any blemishes.

If you are making a piece of furniture such as a bookcase or secretary, you may be using a hardwood. A sash plane will work on a medium-hard wood such as mahogany or walnut if the material is carefully selected for the qualities just described. Otherwise, you might consider making just the frame out of hard wood. These pieces are larger, and you only have to run the plane in one direction. You can use a soft wood for the narrow muntins, and then stain them to match the frame.

Thickness your wood to the width of your plane. If you own a fillister, you can thickness the wood to the dimension you want and adjust the plane to fit. Whether you are making a stile or rail or muntin, always joint the edge.

When making the stiles and rails, clamp the board upright in the side vise as you run the plane down onto the edge of the board. A sash plane is not sprung, so hold the stock vertically. Set the forward end of the sole on the wood, with the fence held tightly against the left side. The mouth should be placed just off the end corner of the board, ready to engage the wood.

Your first pass should cut shavings from both sides of the plane. Each succeeding pass will develop more of the moulding. The only place that is not cut is the area between the irons. In the completed sash, this will be the inside edge of the stiles, rails, and muntins—in other words, the surface closest to the glass. When the shapes are complete, the stop will run on this uncut strip, and the cutting action will cease.

If you are making muntins, clamp the board on the bench top and run the plane on the board's surface. Press the fence against the board's left edge and run the shape. You will notice that the rabbet side now cuts a wide groove, whereas on a stile or rail the cutter makes a rabbet that is open on the right side.

When the shape is fully formed, you have one half of a muntin. Invert the board and run the profile again on the other side. (See Illus. 210.) When you finish that second profile, you will have a completed muntin that is attached to the board by its spline. The spline is the wood that remains between the two rabbets. The muntin can be freed with a fine (7 or more points) ripsaw or even slit with a utility knife.

Joint the edge of the board once again and make another muntin in the same two-step process.

*Illus. 210. Using a sash plane to cut a muntin. The bottom side is already complete. When the second cut is finished, the muntin will be cut free with a 7-point ripsaw. (See Chapter 18.) Normally, the plane is held with two hands. I have dropped my left hand to better show the plane. Note that each side of the plane cuts its own chip.*

# 14
# Metal Special-Purpose Planes

During the second half of the 19th century, while plane-making companies like Stanley and Sargent were perfecting their iron bench planes, they were also developing replacements for wooden special-purpose planes. By the early 20th century, this process was complete, and, like wooden bench planes, wooden special-purpose planes soon ceased to be made.

Many types of metal special-purpose planes have been used by woodworkers for well over 100 years and, as a result, they are common items on the second-hand tool market. However, during the decades when cast-iron planes were being developed, inventors tried numerous alternative designs. Competing designs eliminated many of these attempts. You will occasionally see these obsolete alternative tools for sale. Leave them for tool collectors. The tools described in this chapter have endured the test of time. They gained popularity with woodworkers because they work well.

Many metal special-purpose planes are still being made. Based on my experience, the earlier examples (those made before the mid-20th century) are usually better tools than those made in the last couple of decades. Their manufacturers put more of an emphasis on quality; as a result, their castings have fewer flaws and have better machining. More work went into shaping and finishing wooden parts such as the totes, so they are more comfortable to hold. These parts are often made of an exotic wood, while the parts on some more-recent special-purpose planes are sometimes made of plastic.

The cast-iron bodies of earlier metal special-purpose planes were japanned; black and red were the most common colors. This japanning has usually worn and been chipped over the years, but remains shiny. Around the turn of the 20th century, japanning was sometimes replaced by nickel plating. When new, the nickel plating was bright; however, after several decades it has dulled, and I do not find it as attractive as japanning. However, neither finish affects how well the tool works. Base your purchases on your own aesthetic values.

Before buying a metal special-purpose plane, make sure that it is not missing any parts. Though you may not know exactly how many parts were originally with the tool, such clues as a threaded hole with nothing fitted in it should make you very suspicious.

If the parts on a tool you are considering buying are damaged or missing, you (as a woodworker) probably do not have the skills necessary to either repair or duplicate them. However, these parts are interchangeable and can be salvaged from planes (of the same model) that are damaged or incomplete. In some cases, the manufacturer may still offer replacement parts. For example, the cutters for some of the planes described in this chapter are available through mail-order catalogues.

I still recommend purchasing only complete planes, but if you do want to acquire a particular tool by combining two or more incomplete planes, consider the following: you will spend a lot of time looking to find two of the same tool, especially if it is an unusual type. Your time should be factored into the total price.

Keep in mind the prices you have paid. If you do not have a good memory for such details, keep a written record. If two incomplete planes cost more than a complete one, you have lost money.

Many of the metal special-purpose planes on the second-hand market are still being made. If you are a total pragmatist, the price of a new tool should be the maximum that you're willing to spend for a second-hand one. Take some time to familiarize yourself with new plane prices so that you do not pay more for a second-hand tool.

Almost no new tools are ready to be used as they come from the box. Like old tools, they too have to be tuned and adjusted. The amount of tuning required to return an old plane to working condition and the amount needed to tune a new plane are often about equal.

Most woodworkers want to work with tools that are as fine in quality as the woodworking they do. In this case, you may be willing to pay a little more for an older tool. As I explained in Chapter 1, second-hand tools are often better tools.

# Block Planes

As mentioned in the last chapter, tools with a low cutting angle are intended for use on end grain. Low-angle tools make paring cuts that are much cleaner.

Formerly, cabinetmakers and joiners owned a type of low-angled plane called a mitre plane. In the last chapter I described the wooden mitre plane; for very fine work, cabinetmakers needed a more exact tool. Their mitre planes were often made of metal so that the sole would not wear and the mouth would be tighter. These mitre planes were essentially a metal box filled with an exotic hardwood core. These wood-filled planes were made to such close tolerances that a piece of paper could barely fit through their mouths.

These early mitre planes are wonderful tools to use. They are not rare. However, they are very popular with collectors who are willing to pay high prices for them. If you want to own one, be prepared to spend a sizable sum of money. You will have to contact a specialized tool dealer.

During the late 19th century, the wood-filled mitre plane was superseded by a small cast-iron block plane that is still made today. When properly sharpened and adjusted, these block planes will work nearly as well as their earlier counterparts.

Block planes were made in large numbers. They are very common on the second-hand tool market, so they are not expensive. Of the many types of block planes that were made during the past 100 years, I am only going to include the two that I use. (See Illus. 211.) They are the most common and are both of the best quality. The ones that I have were made by Stanley; both have an 1897 patent date. However, very similar planes were made by several companies at that time.

The bedding angle is the major difference between these two types of block planes. The cutters of these planes are set very low, so that they make clean paring cuts on end grain. One is set at 20 degrees, the other even lower, at 12 degrees. (See Illus. 212.) Both angles are lower than those on a wooden mitre plane; in fact, they are so low that the irons are bedded with the bezels placed upward. If the bezel were placed down as in other planes, its relief angle would be so shallow that the cutting edge would have no strength. Since the bezel is upward, neither cutter has a chip breaker.

Both block planes have a precision longitudinal adjustment. The 20-degree plane has a brass wheel that is mounted horizontally on a left-hand-threaded post. (See Illus. 213.) The wheel connects to a slotted arm, located under the cutter. The end of the arm has two teeth that fit into a row of parallel slots in the cutter's lower surface. As you turn the wheel, it moves the arm, which, in turn, moves the iron. If you turn the wheel clockwise, you advance the cutting edge. It

*Illus. 211. The block plane on the left is a Stanley #15 (20 degrees). On the right is a Stanley #65½ (12 degrees). Both have an 1897 patent date.*

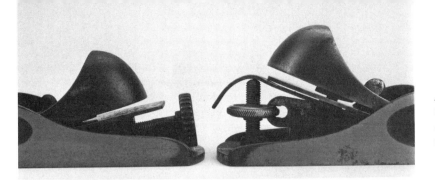

*Illus. 212. When the planes are viewed from the side, the differences between a 12- and 20-degree pitch are obvious.*

*Illus. 213. The cutters have been removed from these two block planes to show the different adjustments. Both types of longitudinal adjustment rely on a tooth that engages one of a series of slots on the cutter's lower surface. The 20-degree plane also has a lateral adjustment. As on a bench plane, the friction-reducing wheel on the end of the lever fits into the slot in the cutter. All these adjustments are very precise and result in planes that can trim the thinnest of shavings.*

*Illus. 214. A pair of block plane cutters. The one with the cut corners is the earlier cutter. It bears Bailey's name and an 1867 patent date. (The one with the round end is more recent and much more common.) Note the row of grooves in the lower surface. The tooth on the longitudinal adjustment engages one of these grooves.*

retracts when the wheel is turned counterclockwise.

The 12-degree block plane has a different mechanism for setting the iron. It also relies on a wheel mounted under the back of the cutter. In this case, the adjustment wheel is iron and is mounted vertically. The wheel is attached to a bolt that screws into a short post. The bolt is unusual in that it has two different diameters and two different sets of threads. The threads on the first half of the bolt have more lead. When the wheel is turned, the first set of threads causes more travel than the second.

An L-shaped arm is screwed to the second set of threads. The arm has a tooth that fits into parallel grooves in the back of the iron. As the wheel is turned, the arm slides in the top of the post, advancing or retracting the cutter.

The iron bench planes described in Chapter 10 have a frog that adjusts the width of the mouth. The block planes described in this chapter make this adjustment via a sliding sole, very much like the one on my transitional smooth plane (see

Chapter 11). The sliding sole is a separate plate fitted into the casting, just in front of the mouth. The plate's rear edge is the front edge of the mouth. (See Illus. 215.)

*Illus. 215. These two block planes have been inverted to show their sliding soles. One sole has been removed.*

To make the adjustment, first loosen the brass knob located on the toe, just in front of the iron. This knob has a threaded stem that fits through a slot in the casting and into a threaded hole in the movable sole. (See Illus. 216.)

*Illus. 216. The toes of two block planes. One has its knob and eccentric removed.*

The stem also fits through a curved slot in a separate device that is called by some a "quadrant," and by others an "eccentric." When you pivot the eccentric, it moves against the stem and pushes the sole forward or back. You secure the mouth's setting by tightening the knob.

Both the 12- and the 20-degree planes use the same adjustments for the mouth, and these parts are usually interchangeable. Both planes also use the same cutter and cap.

Both the cap and the cutter fit over a cap screw. The cap is tightened with a cam lever. It also has a domed end that fits comfortably into the palm of your hand. Otherwise, there is no tote or knob.

The 20-degree plane has a lever for lateral adjustment that pivots on the cap screw. On the lower end of the lever is a wheel that fits into the slot in the cutter. When you move the lever to the left, it advances the cutter's left corner. When the lever on the block plane is pushed to the right, it advances the right corner.

## Selecting and Reconditioning a Plane

Block planes are very common tools on the second-hand market, so do not settle for one that is neither complete nor in perfect condition. A dirty plane can be easily cleaned, and you can remove light rust. Heavy rust may have caused pitting and should be avoided.

Check for the same flaws that occur to bench planes. The casting should not be cracked or badly chipped. Make sure that no parts are missing or damaged. Look at the sliding sole. It should move easily but not sloppily. Remove it to make sure that water has not seeped underneath and caused any damage.

Disassemble the plane and make sure that no parts are badly worn. Check all the threaded parts to determine whether their screws are worn or crossed.

Before using a block plane, disassemble it and clean all the parts. Mineral spirits will remove much of the grime. Make sure that all moving parts are free and move easily. Remove the sliding plate in front of the mouth and clean its bearing surfaces. A silicone spray lubricant such as WD-40 will make the mouth adjustment much easier.

Straighten the plate's rear edge on the lapping table. Jointing this edge will increase the mouth's precision. Return the plate to the plane and joint the entire sole on the lapping table. If you are going to use your plane on a shooting board, also joint the right side and check to be sure that it is square with the sole.

# Sharpening the Iron

To remove the cutter, pivot the cam lever to the right. The cap will slide off the machine bolt, and the cutter can be lifted out the same way.

If the cutter is badly damaged by either rust, excessive grinding, or misuse, buy a replacement. You can remove light rust on the lapping table and correct the edge on a bench grinder.

First, use a try square to check whether or not the cutting edge is at a right angle to the sides of the cutter. If it is not square, spray the side opposite the bezel with layout fluid. Place the side of the cutter against the square's handle and push the edge just slightly above the square's blade. You can now trace a straight edge in the layout fluid with a sharp point such as that on a scratch awl. Grind to this line and make a proper bezel, which should be about 30 degrees.

Flatten and polish the side opposite the bezel. This will also remove rust pitting. Do not forget to hone the cap's front edge to prevent shavings from being trapped under it.

Return the iron to the plane. Place your fingertips over the mouth and rest the cutting edge against them. Whether you are using a 20-degree or a 12-degree plane, make sure that the teeth in the longitudinal adjustment mechanism fit into one of the parallel grooves in the underside of the cutter. Set the cap over the cap screw, making sure that the cam is loosened (to the left).

Use a screwdriver to tighten the screw to a point that the cap is held lightly against the cutter. It should still be loose, but it should not be able to move up and down. Lock the cam. It should move easily and still tighten the cap against the iron. If the cam moves with too much effort, turn the cap screw half a turn counterclockwise. If after tightening the cam you find that the cap is still loose, release it and turn the machine screw clockwise.

Turning the longitudinal adjustment screw, advance the iron until you can feel the projecting edge with your fingertips. You can sight down the sole if you find this technique easier.

If you are using a 20-degree plane, make the necessary lateral adjustment. If you are using a 12-degree plane, you can move the iron with a light hammer. Tap either corner to make the iron move in the direction you need.

Once the iron has been set, adjust the mouth. The size of opening you want in front of the cutter depends on the cutter's set. If it has a heavy set, the mouth will have to be wide enough to allow that thickness of shaving to pass through it. If the cutter has a light setting (this is more common), the mouth should be closed until you can see just a sliver of light when you hold it up to your eyes.

# Using a Plane

The block plane, which is small enough to be held in one hand, is most commonly used for precision trimming. I often use mine by holding the heel against my sternum and pulling the piece of wood across the mouth of the plane with my other hand. (See Illus. 217.)

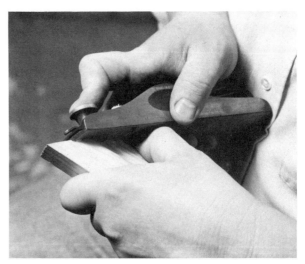

*Illus. 217. I often use my block planes for trimming. This is delicate and exact work. I hold a plane against my sternum and pull the piece of wood (in this case a moulding with a mitred end) over it, removing a paper-thin shaving with each pass. This very precise technique is an excellent way to fit joints.*

The plane will also work on a mitre jack. If you are doing small work, you can build a small shooting board, either for 90- or 45-degree mitres. A block plane is too small for use on a large shooting board.

# Rabbet Planes

Plane factories produced many different types of iron rabbet planes. These planes can be divided into four general types: the shoulder plane, the bench rabbet, the rabbet, and the fillister. Of the four, only the fillister was intended for making rabbets. The other three are primarily used for trimming or squaring.

All four types have one important feature in common—a mouth that extends the width of the sole. As a result, all these planes have open cheeks. However, unlike their wooden counterparts, the mouths of these planes are not skewed.

These planes—and one additional one, the bullnose —are discussed below.

# Iron Shoulder Planes

A wooden version of the shoulder plane (skew rabbet) was discussed in the last chapter. This tool was mainly used by carpenters. Early cabinetmakers and other woodworkers who did precise joinery used a different type of plane that was capable of doing fine work. These precision tools were commonly called shoulder planes. (See Illus. 218.) Like cabinetmakers' mitre planes, these shoulder planes are metal shells filled with wood. Their irons are bedded at about 20 degrees (with the bezel facing upwards) and are held in place with a wooden wedge. This low-angle plane is able to cleanly pare the end grain that occurs in a tenon's shoulder.

When these wood-filled planes were new, they were expensive tools. Because collectors prize them, they are still expensive today. If you are willing to buy one, you will discover that they do an excellent job. These planes are not common on the second-hand market, and you will most likely have to purchase one from a specialized tool dealer.

You can get the same performance from a

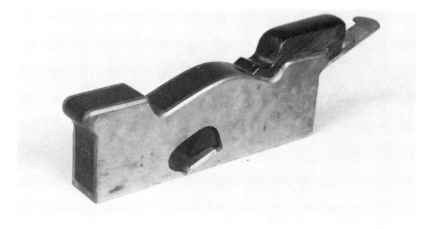

*Illus. 218. A wood-filled shoulder plane made by Norris of London, England. (Photo courtesy of Malcolm MacGregor.)*

*Illus. 219. This Record #311 plane is typical of the low-angled shoulder planes available on the second-hand tool market.*

factory-made iron shoulder plane. Since factories have been making these tools for well over a century, they are not hard to find. When you do find them, they sell for a lot less than the earlier wood-filled versions.

There are many different styles of iron shoulder planes, but they all have similar features and should be tuned in the same way. These planes have a narrow cast-iron body that's between ½ and 1⅛ inches wide. They range in length from about 5½ to 8 inches. (See Illus. 219.)

Like their wood-filled predecessors, these planes have a low-bedding angle, usually 20 degrees, and the bezels of their cutters are placed upwards. In order for the cutting edge to be the same width as the plane's mouth, the cutter is T-shaped.

Unlike wood-filled planes, which use a traditional wedge to secure the cutter, iron shoulder planes have a mechanism for precise longitudinal adjustment, generally a knurled knob just under the cutter's tang. Some shoulder planes can also be adjusted to regulate the width of the mouth.

Many iron shoulder planes have a two-part body that can be separated so that the rear section can be used as a bullnose plane. Bullnose planes have a short sole in front of the cutter. This allows them to work almost all the way into a corner. Bullnose planes will be explained later in this section.

My shoulder plane was made by Record, perhaps during the mid-20th century. Projecting from the back of the body at the same angle as the bed is a threaded rod onto which is screwed a knurled nut with a flange. (See Illus. 220.) The cutter has a horizontal slot cut in its upper end which fits over the flange. As the nut is turned, it moves up or down the rod, its movement either advancing or retracting the cutter.

My plane has a two-part body that separates just in front of the cutter. The front end is attached to the back by a machine screw and is positioned on two locating pins. The nose pulls off the locating pins when the screw is removed. (See Illus. 221.) The manufacturer provided the plane with a shorter nose that could be screwed to the rear section and turn it into a bullnose. The plane was second hand when I obtained it, and this piece was not with the tool. However, this doesn't prevent the plane from working as a shoulder rabbet, and I do own a separate bullnose.

Originally, the mouth was open too much for precision work. I corrected this problem by machining the rear section's mating surface. This reduced the mouth's width, so that now it is difficult to slide a piece of paper through the opening.

The cutter is held in place by a curved cap that has a pivot pin projecting from either side. These pins fit into two short slots in the back of the cheeks. A knurled adjusting knob is under the back of the cap. As the knob is turned clockwise, it presses down on the cutter. This makes it pivot on the two pins and creates pressure along its front edge. This pressure holds down the cutting edge and prevents it from chattering. (See Illus. 222.)

## Bench Rabbet Planes

The bench rabbet was originally called a carriage-maker's rabbet. It also looks very much like a bench plane, which explains its name. (See Illus. 223.) Because these planes are about the same size as a jack plane (12¾ inches, as opposed to 14 inches for the Stanley #5 jack plane), they are even sometimes known as jack rabbets.

Bench rabbets have all the features of an iron bench plane, including a rosewood tote and knob. They have a lateral adjustment lever under the iron and a brass longitudinal adjustment knob in the back of the frog. As on a bench plane, the frog can be adjusted to control the width of the mouth.

Another feature that bench rabbets share with bench planes (but not with shoulder planes) is the bedding angle. The pitch is 45 degrees, just like that on a bench plane. As a result, the cutter is used with the bezel down. However, the iron is

*Illus. 220. To adjust a Record shoulder plane, turn the knurled nut on the end of the threaded rod. The nut has a flange that fits into a slot in the end of the cutter. The adjustment will be very precise.*

*Illus. 221. The #311 disassembled. The front end has two locating pins that ensure proper alignment and can be removed so that the tool can be used as a chisel plane or (with a special attachment that is missing) a bullnose. Note the cap. It has two pins that slide into slots in the cheeks.*

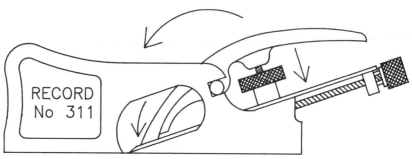

*Illus. 222. A shoulder plane has a screw cap. Once you have adjusted the cutter, turn the thumbscrew. This pushes down on the cutter's tang, which pivots the cap on the two pins that fit into the cheeks and creates pressure on the cutter's front edge. This holds the cutting edge securely against the bed so that the cutter cannot chatter.*

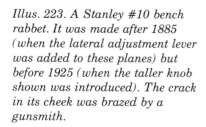

*Illus. 223. A Stanley #10 bench rabbet. It was made after 1885 (when the lateral adjustment lever was added to these planes) but before 1925 (when the taller knob shown was introduced). The crack in its cheek was brazed by a gunsmith.*

also T-shaped and has a T-shaped chip breaker. The cutter/chip breaker assembly is held in place by a cam lever cap, just like on a bench plane.

The high-bedding angle means that the tool will not work as cleanly on end grain, and is more useful for cleaning rabbets that run with the grain. The plane is too large for trimming the shoulders of most tenons. That job is better left to a shoulder rabbet.

# Rabbet Planes

The cast-iron rabbet planes described in this section are used for very much the same purposes

as wooden shoulder planes. Unlike the shoulder planes and bench rabbet planes, iron rabbet planes are not precision tools; they are most useful for carpenters, rather than for joiners and cabinetmakers. (See Illus. 224.)

Iron rabbet planes are quite simple. Those made by most manufacturers look very much alike. They have a low body with a blunt front end, called the toe. They are pushed with a tote that is part of the casting. Unlike shoulder planes and bench rabbet planes, these planes have only a right cheek. The left side of the body is completely open.

The bed on which the cutter is secured is no

*Illus. 224. A Stanley #190 rabbet plane. This one was made after 1912, when the dappled pattern on the handle was introduced.*

*Illus. 225. Because a rabbet plane is open on the left side, it has an unusual asymmetrical cap and cutter. The cap and cutter have a longer right leg that extends through the closed right cheek. Note the bed. It is no more than a platform with a cap screw on it.*

more than an angled platform. A cap screw is threaded into the bed and the cutter and cap are secured over it.

The cutter is a different shape than that on a shoulder plane or bench rabbet plane. (See Illus. 225.) It is a modified T-shape, but the top of the T is offset so that the right side is longer than the left. In other words, the horizontal top of the T is not centered over the tang. The cutter is longer on the right because that side has to extend through the cheek. There is no cheek on the left side, so less length is required. The cap has a similar shape so that it applies even pressure along the entire cutting edge.

The cap has a knurled setscrew in its back end. When this is advanced, it applies pressure on the back of the cutter. At the same time, it is ful-crumed over the cap screw, forcing its front edge against the front of the cutter. The action is very similar to that of the shoulder plane. However, the cap is fulcrumed on a cap screw rather than on pins that fit into the cheeks.

Because the bed is set at 45 degrees, this type of plane is most effective when working with the grain. However, sometimes you have to make rabbets that run across the grain. To help prevent tearout when working in this direction, these metal rabbets usually have a scribe. A similar device is found on wooden fillisters and some wooden rabbets. On these planes, the scribe is usually referred to as a spur.

The scribe on a wooden plane is a long, tapered piece of steel. On an iron rabbet, scribing is done by a small wheel with three spokes (the spurs) around its diameter. (See Illus. 226.) The wheel is set into an inlet of the same shape and is held in place by a flat-headed screw. The inlet is in the outside of the cheek, about ⅜ inch in front of the mouth. Although the wheel has three spokes, the inlet has four. The fourth points downwards and intersects with the sole.

When the plane is not in use or when it is used along the grain, turn the spur wheel so that all its spurs lay in the inlet. When the plane has to be

*Illus. 226. A rabbet plane has a depth gauge on its right side that you can adjust by loosening the thumbscrew. Under the gauge is a spur wheel. When working cross grain, place one spur downward to act as a scribe.*

used across grain, loosen the screw and rotate the wheel so that one of the spurs is directed downwards. In this position, it projects below the sole so that it can function as a scribe and sever wood fibres in advance of the cutter.

On my plane, only one spur is sharpened. Its lower end is becoming worn and in time will become so short that it will no longer be able to do its job. When that occurs, I will sharpen one of the other two spurs. The plane has a 1910 patent date and was actually made some time after that. I am perhaps its third or fourth owner. I doubt that I will ever use up the remaining spurs in my lifetime. Plenty of use will be left for some future owner who may not even be born yet.

Rabbet planes have an adjustable depth gauge very much like those on wooden fillisters. This depth gauge is mounted over the spur wheel and is attached to the cheek with a screw. When the screw is loosened, the gauge can be moved up or down. Tighten the screw to lock the gauge in place.

My rabbet plane was made by Stanley and is a #190. Its sole and cutting edge are 1½ inches wide. Stanley has also made #191 and #192 rabbet planes. The only difference between these planes is the width of their soles. The #191 is 1¼ inches wide, while the #192 is 1 inch wide. From 1886 to 1918, Stanley also made a very similar range of planes called the #180, #181, and #182 rabbet planes. Rabbet planes were also made by Sargent, Record, and other tool companies. Most have a japanned finish that is black or some other dark color. Rabbet planes by any maker are common on the second-hand market, and they are very inexpensive.

# Fillisters

The cast-iron fillister is used for making rabbets. It can also be used in several other ways. For this reason, it is alternately called a duplex, fillister, and rabbet plane, a duplex rabbet plane, or just a duplex plane.

Fillisters were made by several toolmaking companies. Stanley called its version the #78, Sargent referred to its fillister as #79, and Record had the #778. (See Illus. 227.) All these fillisters are very similar and all look very much like the iron rabbet plane described above.

The fillister, like the iron rabbet plane, has a low body with a round, blunt toe. However, the fillister's body is about a half inch longer than the iron rabbet. The body is still open on the left side, but instead of one bed for the cutter there are two. One is in the middle of the body (as on the rabbet plane). The other is in the toe. The sole also has two mouths. The first mouth is in the same location as the mouth on the rabbet; the second mouth has a mere ³⁄₁₆ inch of sole in front of it. The second bed and mouth allow you to move the blade forwards and use the fillister as a bullnose. (See Illus. 228.)

Although the plane has two beds, it only comes with one iron and one cap. Both these parts are very similar to those found on a rabbet plane, and (in the case of Stanley) they are interchangeable between the rabbet and the fillister.

Like the rabbet plane, both of the fillister's beds have a cap screw over which the cap and iron are mounted. The earliest fillisters have neither longitudinal nor lateral adjustment. In 1925, Stanley introduced a longitudinal adjustment lever

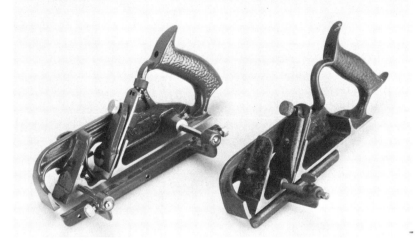

*Illus. 227. A pair of fillisters. The one on the right is a Stanley #78 that was made after 1909, when the dappled pattern on the handle replaced the earlier scroll. A more recent Record #778 is on the left. Note its longitudinal adjustment, which is the same as that on the Record shoulder plane. Also, the fence on the Record plane has two rods, instead of one.*

*Illus. 228. The fillister arranged as a bullnose.*

that was placed behind the first bed. Another method for advancing and retracting the cutter is found on Record's fillister. This is the same device used on some Record shoulder rabbets, and which is described above. It consists of a threaded rod projecting from behind the bed. Mounted on the rod is a round nut with a flange which fits into a horizontal slot in the upper end of the blade. As the nut is threaded up or down the rod, the flange advances or retracts the cutter.

The fillister also has the same sort of depth gauge as the rabbet plane. It also has the same type of scribe wheel mounted in front of its first mouth. The second mouth does not have a spur wheel, as you do not usually need a scribe when the plane is used as a bullnose.

Like its wooden counterpart, the iron fillister has a movable fence. This allows it to make rabbets of any width up to a maximum of 1½ inches. The fence is also made of cast iron and is mounted on a steel rod that is threaded into the side of the body. The fence slides on the rod and is fixed in position by a setscrew. There is a threaded hole in either side of the casting so that the fence can be mounted on both sides of the plane. I have never mounted mine on the right side, and suspect that this feature is not very practical.

There are some slight changes in later-model fillisters made by Record. The fence is mounted on two rods, one in front of the first mouth and the other in front of the tote. These holes pierce the entire body so that the rods can be slid back and forth through them. To secure the rods in place, tighten the setscrew in the body. To set the fence on the rods, also tighten its setscrews.

## Bullnose Planes

In describing the shoulder plane, I noted that some of these tools could be converted into a bullnose by removing the front of the plane and screwing on a special short sole. Other planes were made specifically as bullnoses.

A bullnose plane is related to rabbet planes, but

has a specific use. It has a very short sole in front of the mouth (about ³⁄₁₆ inch) so that the tool can reach almost all the way into a blind corner. The short distance that cannot be reached by the bullnose plane can be easily cleaned up with a chisel.

I own and use a Stanley #75, a bullnose plane the company started making in 1879 and still makes today. (See Illus. 229.) This plane is similar to those that were made by other tool companies. These bullnoses are common on the second-hand market and are usually quite inexpensive.

*Illus. 229. A Stanley #75 bullnose plane.*

Most bullnose planes (but not all) have short, two-part bodies. The lower section holds the iron, while the upper one has two open, curved cheeks that arch down in front of the cutter. The front end of this section creates a short length of sole.

The two parts are usually held together by a screw that passes through a slot in the back of the upper section and is threaded into the lower. When you loosen the screw, the top section can be moved the length of the slot. This movement allows you to adjust the width of the mouth.

On some bullnoses, the upper section can be removed and the tool used as a chisel plane. In this configuration, the open blade is able to reach all the way into a corner. However, I've found that these planes, called chisel planes, do not work well.

Some bullnose planes have a longitudinal adjustment mechanism similar to that described on the shoulder rabbet. On the less elaborate #75 planes, adjustment is done with your fingers. The #75's blade is held in place by a cap similar to that on the fillister. (See Illus. 230.) The cap is slid under two tabs, one on the inside of each cheek. On the back of the cap is a knurled setscrew which, when advanced, tightens the cap. As the screw advances, the rear end of the cap is lifted so that the cap fulcrums on the tabs. This causes its front edge to press against the lower end of the cutter.

The sides of some bullnoses are flush and at right angles to their soles. The #75 is different. Both sides of the lower section have a narrow projecting surface (about ⁵⁄₁₆ inch high) that is flush with the edge of the cutter and square with the sole. These surfaces ride against the side of the rabbet. Above them, the sides of the plane become narrower so that the edges of the cutter project slightly. Because of these features, the #75 would not be a good tool for making or cleaning rabbets. These jobs are done by fillisters and rabbet planes. A bullnose has a limited purpose, and only works when used in corners.

Like shoulder planes, the irons of some bullnose planes are bedded at 20 degrees. Their bezels are placed upwards. The #75's cutter is bedded at 45 degrees, so that the bezel is placed downwards. The bodies of some bullnoses are nickel-plated, while others, like the #75, are japanned black or some other dark color.

# Selecting and Reconditioning a Plane

Shoulder planes and bench rabbets are precision tools, made for trimming the shoulders on rabbets and tenons. Therefore, you should not buy a plane that has been damaged, abused, or is heavily worn. The requirements for an iron rabbet, fillister, or bullnose are not as rigid.

Any of these planes you might buy—with one exception—should still have all of its parts. I cannot use my shoulder plane as a bullnose because it is missing its special short sole. However, it still works very well as a shoulder plane, and I do have a separate bullnose plane. Some minor parts, such as the longitudinal adjustment knob on the bench rabbet, are interchangeable with bench planes. However, important parts like the frog are not.

Fillisters and #190-type rabbet planes are common on the second-hand tool market and are usually priced lower than new planes. However, many are missing their depth gauges. Also, make sure that the spur wheel is still in place. Some

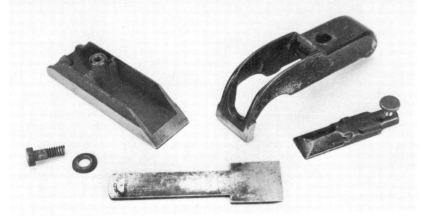

*Illus. 230. A bullnose plane disassembled.*

fillisters are sold as rabbet planes because they have lost their fences. Look for the holes for the rods.

The sides of any rabbet plane have to be at right angles to the sole. Otherwise, the shoulders of the rabbet (or a tenon) will not be square, and the joint will not fit well. Before purchasing any one of these planes, test it with a try square. Although some improvements can be made on the lapping table, do not buy a rabbet plane that is not at least close to square.

The castings on rabbet planes are weak above the open cheeks. Examine the castings very carefully for cracks, as they are very difficult to repair. My bench rabbet had a broken casting on one side, and the only person I could find who had the skill to braze it was a gunsmith.

The first step in tuning any type of rabbet plane, fillister or bullnose is to disassemble it and clean its parts. Mineral spirits will remove much of the dirt and grime.

The wide end of a T-shaped cutter is the same width as the plane's mouth. The T-end projects so that it is flush with the outside edge of the open cheeks. This makes it too wide to be removed by drawing it straight back. It catches in the cheeks.

To remove the cutter on a shoulder plane you have to raise it off the bed and carefully roll it 90 degrees. On a rabbet plane or a fillister, loosen the setscrew and draw the cap back as far as the slot will permit. Push the upper end of the cap to the right so that it turns on the cap screw. (See Illus. 231.) This will withdraw the long end from the cheek. The cap, and then the cutter, can then be lifted off the screw.

The procedure used to remove the cutter on a shoulder rabbet will not work on a bench rabbet. Its double iron makes it too thick to be able to roll out of the cheeks. To remove the cutter, first

*Illus. 231. To remove the cap from a rabbet or a duplex plane, loosen the setscrew. Next, push the cap's upper end to the right. This pulls the cap's leg free from the cheek. It now can be pulled rearward and lifted off the cap screw.*

remove the cap screw that anchors the cutter/chip breaker assembly to the frog. Next, remove the screw that holds the cutter and chip breaker together. Lift the chip breaker and hold it out of the way while you slide the cutter out of the mouth. (See Illus. 232.) Now, the chip breaker is able to slide out the same way.

Once the iron is out of your plane, you can remove the frog and clean it. Also clean the surfaces on which the frog is seated. Obviously, removing the iron requires so much effort that you will not want to do it often. It is an incentive to take good care of your bench rabbet.

Loosen the cap on a #75-type bullnose by turning the setscrew counterclockwise. Remove the cap by pulling it directly back. As on a bench rabbet, you can remove the #75's cutter by pushing it out of the mouth. (See Illus. 233.)

While the plane (whether a rabbet, bench rabbet, shoulder, fillister or bullnose) is disassembled, joint the sole on the lapping table. A plane with a two-part body, such as a shoulder or bullnose plane, should be put back together before jointing.

# Sharpening the Iron

These planes do not have skewed irons, so the iron's cutting edge has to be square with its sides. This is critical, as the sides of the cutter also have to be flush with the open cheeks. If the cutting edge is not square, you can adjust it laterally, but

*Illus. 232. Remove the cutter from a bench rabbet by separating the cap and cutter. Then, lower the cutter out of the mouth.*

*Illus. 233. The cutter on a bullnose plane is also removed through the mouth.*

then the sides of the cutter may project beyond the cheek. When the plane is used, the cutter's projecting edge may lift the sole so that the tool trims the rabbet out of square. It could also damage your woodworking. (See Illus. 234.)

Use a try square to determine whether or not the cutting edge is square with the sides of the cutter. If it is not at a right angle, spray the side opposite the bezel with layout fluid and use a try square to trace a square line. Grind a new edge to this line.

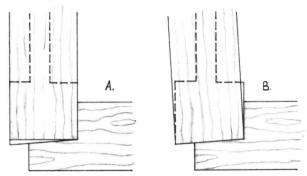

*Illus. 234. The cutting edge of a rabbet plane iron must be perfectly square with the right side. If not, you will have great difficulty making a square rabbet. The cutter shown here is out of square. In A, its right side is set so that it is flush with the side of the plane, but the edge is not parallel to the sole. As a result, the rabbet is out of square. In B, the cutting edge is parallel to the sole, but now the right side projects beyond the cheek. As a result, it rubs against the rabbet's shoulder, pushing the plane out of square.*

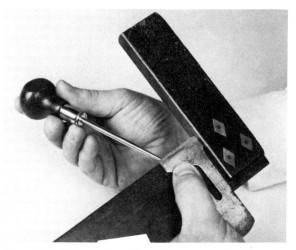

*Illus. 235. The cutting edge of any type of rabbet plane must be square with the sides. Check with a try square. If it is not, trace a line with layout fluid and grind to that line.*

Flatten and polish the surface opposite the bezel and hone the edge as you would that on a bench plane iron. The bench rabbet has a double iron, so remember to hone the front edge of the chip breaker as you would that on a bench plane.

Put the cutter back into the plane in the same way it was removed. To set the cutter of any of these planes, hold the body so that your thumb and index finger are placed at the corners where the mouth meets the open cheeks. This pressure holds the cutter in position. (See Illus. 236.)

Install the cap and tighten the setscrew until it creates just enough pressure to hold the iron in place. You can now do any necessary lateral adjustment without the iron shifting longitudinally. When you see the cutting edge just barely projecting out the mouth, you should have enough set. Finish tightening the setscrew so that the cap will hold this setting while the plane is in use.

On a bench rabbet, adjust the frog before returning the cutter to the plane. You cannot know the correct position until the iron is in place. This means that later on you may have to take the cutter out and make more adjustments. Start by making your best estimate the first time.

You have to return the cutter to a bench rabbet by sliding it up the mouth. The cutter is preceded by the chip breaker; assemble the two while they rest together on the frog. Hold the two in position by gripping them through the cheeks with your thumb and middle finger. Your index finger can keep them from sliding back out the mouth. Remember to place the chip breaker's edge about 1/16 inch behind the cutting edge.

Pass the cap screw through the slot in the chip breaker/cutter assembly and screw it into the frog. Install the cap as you would the cap on a bench plane. The cam should be tight, but it should not require too much effort for you to loosen it. Adjust the plane longitudinally by turning the brass nut in back of the frog. When the cutting edge projects out of the mouth, sight down the sole to make sure that it is parallel. If not, make minor lateral adjustments, making sure that the edges of the cutter do not project beyond the cheeks.

# Using a Plane

To trim a rabbet with a rabbet plane, shoulder plane, or bench rabbet, hold the side of the plane firmly and evenly against the shoulder. The cutting edge should be just off the edge of the work so that as you push the plane it engages the wood. Make a complete unbroken pass. (See Illus. 237–240.)

When trimming tenon shoulders with a shoulder plane, use the plane in short strokes. Do not run the plane across the far corner, as this risks chipping it. (See Illus. 241 and 242.)

Always hold a bullnose tightly against the side of a rabbet and, with short strokes, work into a corner. Do not try to lift the plane even though it leaves a chip still attached. You generally have to clean up the last 1/2 inch or so with a chisel.

Before using a fillister, determine the width and depth of the rabbet you need to make. Set the depth gauge and then the width of the fence. Before making the rabbet, joint the board and

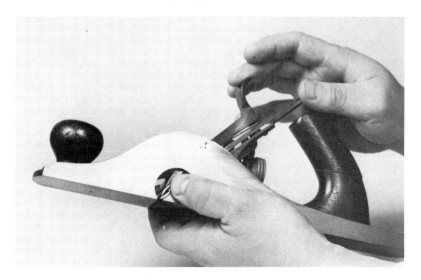

*Illus. 236. When you return a rabbet plane cutter to the plane, hold it in place by gripping its edges through the open cheeks.*

*Illus. 237. Using a bench rabbet. Two hands are required. I have dropped my left hand to better show the plane.*

*Illus. 238. Using a bullnose plane to clean a blind rabbet.*

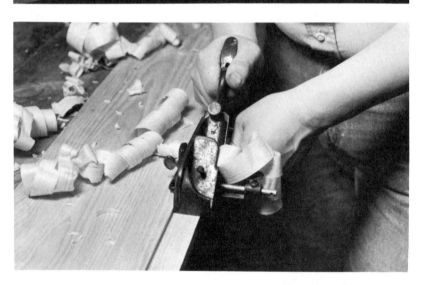

*Illus. 239. Using a duplex plane as a fillister.*

*Illus. 240. Using a rabbet plane.*

158

*Illus. 241. Using a shoulder plane to trim a tenon's shoulders. This plane is so accurate that it pares dust. The result is a very precise fit.*

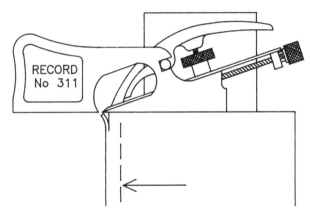

*Illus. 242. When using a shoulder plane to square a tenon's shoulder, do not trim all the way across the shoulder, as this will chip the far edge. Trim to approximately the dotted line shown here. Reverse the plane and finish the far corner from the opposite direction.*

then clamp it to the edge of the workbench. I usually use a tail vise and bench dogs.

If you are using the plane across grain, be sure to turn the spur wheel so that its sharpened tip is pointed downwards.

# Side Rabbet Planes

There are only several planes with iron versions that are decidedly better than their wooden counterparts. This group would include block planes, shoulder planes, compass planes, and side rabbets. As discussed in the last chapter, side rabbets are used for trimming the edges of dados, rabbets, and grooves. Wooden side rabbets come in pairs that consist of a left and a right rabbet; this allows you to always follow the grain, even in highly figured woods.

Stanley developed an all-metal side rabbet very late in the 19th century. Like their wooden antecedents, these metal side rabbets were made in pairs. Another version of the metal side rabbet plane was developed in the 1920's. This was the Stanley #79, and it combined both a left- and a right-hand cutter in a single tool. These planes are still being made today, and are essentially unchanged. (See Illus. 243.) Record developed a different version of the two-in-one metal side rabbet.

Pairs of metal side rabbets stopped being made around World War II and, although they are sometimes found on the second-hand market, the single tool is by far the most common. I use a Stanley #79, and it is the one I will describe in this section.

The plane has a narrow, nickel-plated cast-iron body only five inches long. If the plane were bisected vertically, you would note that the ends are mirror images. Thus, I need only describe one end.

Metal side rabbets, like the wooden versions, have a vertical sole that is flush with the back of the cast-iron body. (See Illus. 244.) The mouth is a vertical slot less than 7/8 inch from the lower corner. The upper corners have a raised shoulder that fits into your palm to help prevent blisters.

The iron is set into a relieved bed that is machined into the casting. The bed is pitched about 12 degrees to the flat back of the body and 30 degrees to the lower edge. The cutter is bedded with its bezel facing away from the sole.

The low bedding angle indicates that the tool is designed for use on end grain, which it must cut when trimming either a dado or a rabbet that runs across the grain. The cutter is held in its bed by a cap. A setscrew passes through the cap and is threaded into a hole in the body. There is no adjustment mechanism, and you have to set the iron with your fingers.

The corner of the sole, the part in front of the cutter, is removable. You only have to loosen the

Illus. 243. A Stanley #79 side rabbet. Earlier side rabbets (available until 1941) came in pairs. This plane (introduced in 1926) has the left and right cutters built into a single tool. The cutter and cap have been removed to show the bed and throat. The nose piece can also be removed to create a side rabbet chisel plane that will reach into a corner.

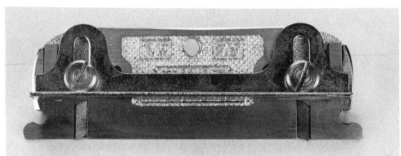

Illus. 244. The back surface of a side rabbet is technically considered the sole. Later-model #79's have a fence that you can adjust by loosening two setscrews.

machine screw that is on the same side as the sole. Removing this piece allows the side rabbet to trim into the corner of a stop dado (one with a blind end) or the corner of a rabbet (such as the top of a doorjamb).

Later models (made after the mid-20th century) were provided with a fence. This is a piece of stamped steel bent to a right angle and attached to the back of the casting by two setscrews. It is adjustable and can be moved so as to cover more, or less, of the mouth. The fence is set to the same depth as the dado, groove, or rabbet you are trimming. As long as its horizontal sole is held on the surface of the board, it ensures that the side rabbet is cutting at a right angle.

The cutter is made of ⅛-inch tool steel that is bevelled along its lower edge. This edge, and the cutter's skewed bezel, meet at the lower side at a 60-degree angle.

## Selecting a Plane

Do not buy a plane that has a damaged casting. Missing parts, such as the cutter caps and the removable corners, cannot be easily replaced. On the other hand, the screws that hold the parts (cap, the removable corners, and the fence) are standard and can be purchased at a hardware dealer. Replacement cutters are still available from tool catalogues. The fence is a handy device,

but not absolutely essential, and the tool will work without it. You just have to be more careful to hold it perfectly vertical.

Before using a side rabbet, take a moment to joint the sole on the lapping table. The sole is really the lower edge of the flat back, but cannot be treated as a separate surface. Flatten the entire back of the casting.

## Sharpening the Iron

Check the cutter's skewed bezel. If it has been sharpened many times, it may no longer have the correct angle (60 degrees). If not, one corner of the cutting edge will project more than the other. When the iron has a light setting, only one corner of the edge will engage the wood and you will not trim the entire surface. When the setting is heavy enough to trim the entire shoulder, the thickness of the cut will be uneven. Both circumstances make it very difficult to trim the sides of the dado at a perfect right angle.

To correct the skew angle, spray layout fluid on the surface opposite the bezel. Hold the cutter firmly against the bed. Also hold the cutter's lower edge tightly against the bed's lower edge. If it moves, the angle will be wrong. Advance the cutter until its edge peeks beyond the edge of the mouth. With a stylus, such as a scratch awl (or in a tight space, a sewing needle), trace the line of the mouth in the dried fluid. (See Illus. 245.) Regrind the edge to this line.

Before honing either an original cutter or a replacement, begin by flattening and polishing

*Illus. 245. To establish the correct angle on a side rabbet cutter, use the edge of the mouth to trace the line.*

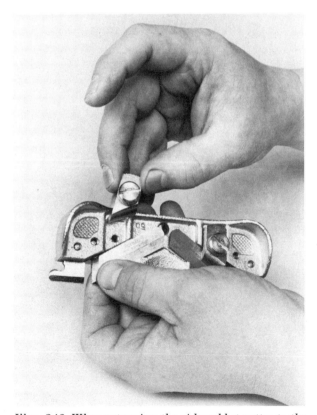

*Illus. 246. When returning the side rabbet cutter to the plane, hold it in the correct position with your thumb. With your free hand, place the cap over the cutter and tighten the setscrew.*

the back. Replacement cutters are coarsely ground and have a pronounced wire edge. Remember, the bezel of a side rabbet is skewed. When honing it, make sure that you apply even pressure along the entire cutting edge. Uneven pressure will change the 60-degree skew and result in one end of the cutting edge projecting more than the other.

Return the cutter to the plane, laying it in the bed, bezel up. Turn the setscrew to the point where the cap begins to tighten (you should still be able to slide the cutter with your thumb). Push the cutter firmly against the bed's lower edge while simultaneously pushing it forward, so that the cutting edge slides into the mouth. (See Illus. 246.) Test the setting by running the tip of your finger over the mouth. When your finger scrapes the cutting edge, the setting is nearly correct. Tighten the setscrew and test the set on the side of a dado, rabbet, or groove. If it is too light, repeat the process, advancing the edge slightly. If the set is too heavy, repeat the process; only this time retract the cutter slightly.

## Using a Plane

If your plane has a fence, lay the tool in the dado, rabbet or groove until the lower edge of the casting is resting on the bottom. Lower the fence until it makes contact with the surface of the board and tighten its setscrews. As you trim, make sure that the fence lays flat on the adjoining surface. (See Illus. 247.) If your plane does not have a fence, be very careful to keep it perfectly vertical.

A side rabbet is also handy for trimming sliding dovetails. These joints are often used in carcass construction to attach the top and the dividers of a chest of drawers and to hold the shelves in bookcases. For example, on some chests of drawers the top is attached with sliding dovetails. Each side has a long dovetail tenon on its upper end that fits into a dovetail-shaped dado in the underside of the top. When in place, the top cannot be lifted off the side, but the sliding dovetail accommodates the seasonal movement of both surfaces. (See Illus. 248.)

To trim either a dovetail tenon or its dado, remove your plane's fence. This will allow the plane to be tilted at a slight angle. (See Illus. 249.)

*Illus. 247. Using a side rabbet plane to trim a dado's shoulders. The plane shown here is a Stanley #79. Its cutters are bedded at such a low angle that they pare tiny shavings.*

*Illus. 248. Side rabbet planes are handy for making sliding dovetail joints. These joints are excellent to use when you are working with wide surfaces (in this case, the top and sides of a chest of drawers), as they accommodate the seasonal movement of wood. The arrows indicate the shrinkage and expansion of the sides and top.*

*Never use glue in a sliding dovetail. The glue constricts the movement of the wood, and even when the joint is fit together dry, it is exceptionally strong. The top cannot be pulled off the long dovetailed tenon. In fact, a chest made this way can be lifted by its top. Note that it is only necessary to taper one side of the tenon and groove.*

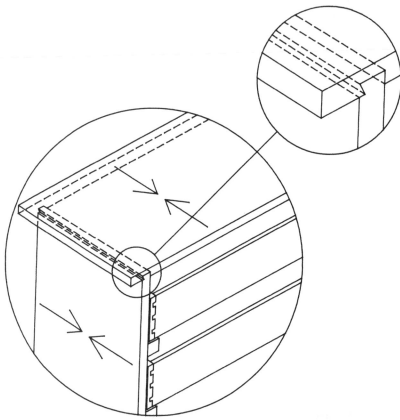

*Illus. 249. A metal side rabbet can be used to make sliding dovetails. First, cut a dado and use the plane to bevel the sides.*

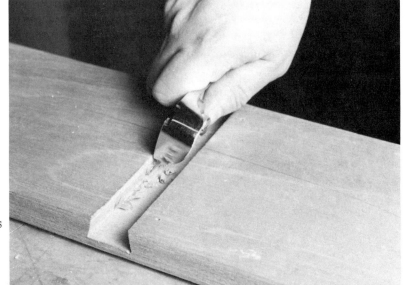

# Dado Planes

From 1902 until 1952, Stanley made a series of cast-iron dado planes. (See Illus. 250.) I do not know if any other companies produced them. Like their wooden counterparts, the depth these metal dado planes cut is adjustable, but the width is fixed (their widths are the same sizes as their soles). Stanley offered eight different sizes. They are, as follow: ¼, ⅜, ½, ⅝, ¾, ¹³⁄₁₆, ⅞ and 1 inch. All these planes are referred to as #39's. However, they are usually listed by dealers as follow: #39, ⅜; #39, ½; #39, ⅞; etc. Stanley has also cast the size into the body of the plane, for example, "No 39. ½ in."

Iron dado planes have all the features of wooden dado planes, although they also share the characteristics of some other metal special-purpose planes. A depth gauge is in front of the iron on the left side. The gauge is attached to an adjustment screw that's threaded through the top of the casting. The screw is similar to that on the 12-degree block plane. It has two sets of threads—one fine, and the other coarse. (See Illus. 251.) The (low-pitch) fine thread is screwed into the casting; the coarse (steep-pitch) thread is threaded through a boss in the back of the depth gauge. Because the fine thread has less lead than the coarse thread, one turn of the screw makes the gauge travel a greater distance than the screw travels in the hole in the casting.

Once the depth gauge has been set to the desired depth, it is locked into position by a thumbscrew. A similar arrangement is used to secure the depth gauge on the fillister and the rabbet plane.

In front of the mouth are two scribes, one on each side of the body. These are vertical blades, rather than the spur wheel seen on the rabbet plane and fillister. The scribes are set into shallow, vertical inlets and are themselves adjusted for depth when you loosen two screws with a screwdriver. A dado plane is usually used cross grain and, as on a wooden dado, these scribes decrease tear-out by snipping the wood's fibres in advance of the cutter.

The plane's iron is set askew. (See Illus. 252.) A skewed blade is found on most wooden planes that are used cross grain, but is unusual on a metal plane. The cutter is secured in place by a cap that is fastened to the body by a screw, on which it also pivots. There is a setscrew at the upper end of the cap. As this setscrew is turned, it applies pressure to the back of the cutter. This pressure causes the cap to fulcrum and apply equal pressure to the front of the cutter. To remove or adjust the blade, simply loosen this setscrew.

The iron itself looks like that on a wooden dado plane; it is trapezoid in shape and is located at the end of a long, thin tang. (See Illus. 253.) Because the mouth is set askew, the iron's cutting edge is at an angle to its sides. So that these sides do not project beyond the cheeks, they are also bevelled relative to the cutter's upper and lower surfaces.

The body itself looks like that of a narrow rabbet plane, with the handle an integral part of the casting. The body's right side is ground smooth and at a right angle to the sole. The left side is open, like the left side on the fillister. The toe is blunt and slightly rounded at the upper corner. Dado planes were usually japanned in black.

Dado planes are not common tools, but neither are they rare. They are generally found in specialized tool shops, or can be purchased through second-hand tool catalogues. They are usually priced slightly higher than a fillister, but are less expensive than you would assume given their limited availability. The exception is the #39,

*Illus. 250. Stanley #39 dado planes. The upper right plane has been reversed to show the right side.*

*Illus. 251. To adjust the depth gauge, first loosen the thumbscrew. Next, turn the adjustment screw. It has a coarse and fine thread so that the depth gauge travels faster than the screw does through the casting. Note the scribe inletted into the side of the casting. Its mate can be seen on the far side of the sole.*

*Illus. 252. The sole of a #39 dado plane. Like their wooden counterparts, these tools have skewed mouths. Note the depth gauge and scribes.*

*Illus. 253. The cutters of #39 dado planes. These irons measure ½, ⅞ and ⅜ inch. The ⅜-inch iron is so narrow that its cutter has to be offset from the tang.*

¹³⁄₁₆-inch dado plane which is a rare collector's item and can be very expensive.

It is doubtful that you will ever need all eight sizes of dado planes, and can probably do all you need with three: a small, a medium, and a large plane. You can then dimension your wood to fit your nearest sized plane. Since one-inch lumber is now thicknessed to ¾ inch, you might find that size more handy than ⅞ or 1 inch.

Before using a newly acquired dado plane, disassemble it and clean the parts. Joint the sole on the lapping table, making sure that you first remove the scribes. If there is any rust on the right side, joint that as well. When you reassem-

ble the plane, make sure that the adjustment screw on the depth gauge is clean. Also clean its threaded holes in the body and in the gauge itself.

## Selecting a Plane

Even though dado planes are unusual tools (normally found in specialized shops), their prices are so reasonable that I would not buy one that is less than perfect. Their parts are not interchangeable with most other iron planes, and most of the parts have complicated shapes. It would take more time to make a replacement than the plane is worth. Do not buy an incomplete plane either, unless you already owned another incomplete one that has the missing parts. Since you often have to buy these planes through the mail, have the dealer assure you that any dado plane you are receiving is complete and undamaged.

## Sharpening the Iron

The cutter of an iron dado plane is sharpened in exactly the same way as that on a wooden dado plane. Follow the advice given in Chapter 12.

Set the iron on a dado plane as you would the iron on a rabbet plane. As on rabbet planes, there is no lateral or longitudinal adjustment mechanism. While putting the iron in the plane, place your thumb and index finger at the corners of the mouth. (See Illus. 254.) This holds the cutter in position while you tighten the setscrew at the back of the cap.

If you have to refine the adjustment, loosen the setscrew slightly, and, while sighting down the sole, move the cutter with your fingers. Retighten the setscrew before using the plane.

## Using Dado Planes

Iron dado planes are used the same ways as their wooden counterparts. (See Chapter 12 and Illus. 255.)

# Match Planes

As mentioned, some wooden match planes were made in pairs; others combined both the tongue and the groove functions into a single tool. All metal match planes will make both the tongue and groove. Some planes have two separate sides and two separate totes. There are two separate cutters, so that the tool can be used one way for one function and can then be reversed and used the opposite way for the other function.

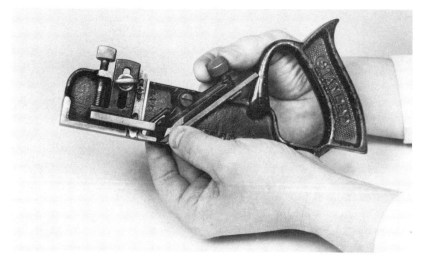

*Illus. 254. When returning the dado cutter to the plane, hold it in place with your thumb and forefinger while you tighten the cap.*

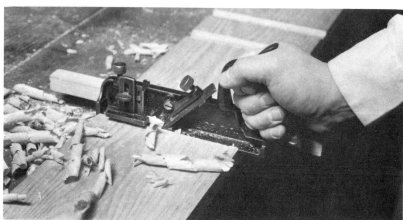

*Illus. 255. Using a Stanley #39 ⅞-inch dado plane.*

The type of plane I own has an ingenious pivoting fence that changes the function. This tool is by far the most common iron match plane on the second-hand market. (See Illus. 256.)

*Illus. 256. This Stanley #48 match plane is japanned and probably dates back to before 1898, when #48 match planes began to be nickel-plated. Stanley made a large number of these planes between 1875 and 1944. They were also produced by several other manufacturers, and are a common item on the second-hand tool market. The right-hand cutter on this plane is longer than the original one because it was taken from a #45 combination plane.*

Wooden match planes were made in graduated sets, that is, pairs of planes were sized for specific thicknesses of lumber. In Chapter 13, I also explained that in an emergency, you could use a slightly oversized or slightly undersized plane. However, the tongue and groove would not be centered on the edge.

Iron match planes were also made in different sizes. Instead of five graduations (as in wooden match planes), there were only two of the type I will describe. Each plane was intended for use on more than one thickness of wood. For example, the plane that centers on ⅞-inch stock will work on lumber between ¾ and 1¼ inch. The plane that centers on ½ inch stock will work on lumber between ⅜ and ¾ inch.

The iron match plane has a narrow cast-iron body. On the front is a turned rosewood knob like those on contemporary bench planes. The tote is part of the casting and is decorated with a pattern of embossed, intertwining scrolls. The sole is divided, front to back, by a ⅜-inch-wide groove.

On the left side of the stock is a fence. It is attached to the plane by a screw that fits through a boss. The pivot is offset so that when the fence is turned 180 degrees the inside surface is shifted. A steel, mushroom-top locking pin is in the front left corner of the stock, just in front of the knob. The pin is mounted in two small bosses and is loaded by a small coiled spring. (See Illus. 257.)

*Illus. 257. To adjust the fence on a #48 or #49 match plane, first lift the locking pin.*

The plane uses two separate cutters made of ¼-inch-thick steel. They are 5/16 inch wide and 3 inches long. Originally, the plane had a third cutter that was about twice as wide as the other two. I have never seen an iron match plane for sale that still had its third cutter. I will explain the use of this cutter later.

The match plane has a 45-degree pitch. Its

cutters rest on two separate beds, one on either side of the stock. The cutters are secured on the beds by two caps that are mounted on a common axis. On the back of each cap is a screw with a knurled brass knob. When tightened, the screw pushes against the back of the cutter. The cap pivots on the axis and applies equal pressure to the front and back of the cutter, holding it securely on the bed.

When the two cutters are in place, look at the sole. It is easy to see how the tool cuts a tongue, as the cutters are separated by a ⅜-inch gap in the sole. As on the wooden version, the upper surface of this gap is the stop.

It is more difficult to imagine how the plane cuts a groove. To set it for this function, you have to turn the fence around. Lift the locking pin; this will free the fence so that it can pivot. Turn it 180 degrees, at which point the pin will line up with another hole in the fence.

Now, look again at the bottom of the plane. Since the fence is offset, the inner edge has shifted and now covers the left-hand cutter. The right-hand cutter is centered to make the groove.

Unlock the fence and pivot it back to the tongue position. If you make the tongue on lumber that is thicker than ⅞ inch, the right-hand cutter will not reach the far edge. You will in effect cut two tongues, the one you need (which will be shifted toward you) and a thin strip that remains on the far corner of the edge. (See Illus. 260.) You can remove the unwanted tongue with a rabbet plane, but this is a waste of time and effort.

The plane was originally equipped with a third cutter that was extra wide, ⅝ inch. It was to be installed in the right side of the plane when the tongue was being made on wider stock. Its extra width would extend over the far corner and eliminate this unwanted, second tongue. You cannot use the wide cutter when making the groove.

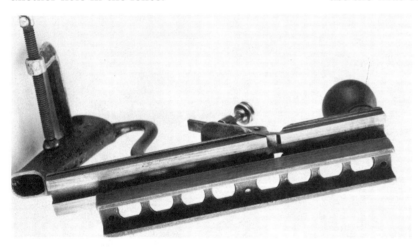

*Illus. 258. The #48 arranged as a groove plane. The darker space between the soles is the stop, which works when the fence is in either position. The plane shown here is held upright by a C-clamp so that it could be photographed more clearly.*

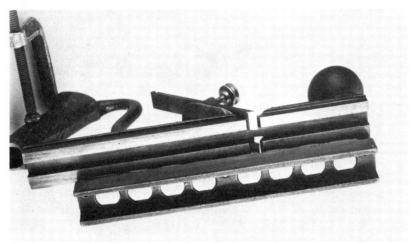

*Illus. 259. The #48 arranged as a tongue plane. Note the wide cutter.*

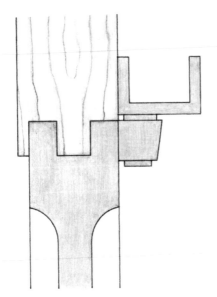

*Illus. 260. If you use a ⅞-inch match plane on a thicker piece of wood, you must use an extra wide cutter on the right side. If not, as shown here, you will create a second tongue.*

## Selecting a Plane

Match planes are made almost entirely of cast iron and steel. The exceptions are the rosewood front knob and the brass tops on the setscrews. It is very difficult to repair cast iron, so avoid a plane that is damaged or heavily rusted.

Check all the parts and sight down the sole and fence to make sure that they are straight. All moving parts should move freely, and the plane should retain at least some of its finish. All the parts should be with the plane. The tool is no longer being manufactured, so you cannot call the factory and order a replacement.

These planes are common. Shortly before writing this chapter, I spent a Sunday afternoon visiting flea markets and antique shops. I saw three of these planes during that day. Since the tool is easy to find, you can salvage missing parts from another match plane. But since complete planes are so common, it does not make much sense to take this approach.

## Sharpening the Iron

I have examined many of these iron match planes and have never seen one that still retained its wide cutter. Fortunately, a ⅝-inch rabbet cutter from a combination plane (see Chapter 15) will work very well. An alternative is to make this

part, which is so simple it can be easily made. You will need a piece of ¼-inch-thick steel (01 will do fine) that is three inches long and ⅝ inch wide. You can make the smaller cutters just as easily. Follow the advice given in Chapter 6 when heat-treating a new cutter.

The bezel has to be at a right angle to the sides. Test the edge with a small try square. Before honing, do not forget to flatten and polish the side opposite the bezel.

## Using a Plane

Use an iron match plane as you would a wooden one. Clamp the board with the edge up in the side vise of your workbench. If you are making the tongue on a thick piece of wood, mount your wide cutter on the right side of the plane (the side towards the bench). (See Illus. 261.) Place the fence against the surface of the board closest to you. Identify the inside and outside surfaces on every board and be careful to always run the fence against the same surface when making both the tongue and the groove.

*Illus. 261. Using a #48 to make a tongue. The plane is cutting two chips; one is partially obscured by the knob. Normally I would be holding the knob, but here I have dropped my left hand for the photograph.*

Hold the plane upright, your right hand on the tote and your left on the wooden knob. Push the tool forward. When you use it to make the tongue, it will cut two shavings. If you are using the ⅝-inch cutter on extra-thick stock, the right-hand shaving (nearest the bench) will be wider than the one on the left. You cut one shaving when making the groove; you must cut it with one of the smaller cutters. The tongue is finished when the plane's stop rides on either the top of the tongue or the top right side of the groove.

When one side is finished, turn the board end over end. The other edge is now up, but the same surface faces away from the bench. Pull up the locating pin to free the fence and pivot it 180 degrees. It is now in position to make the other part of a tongue-and-groove joint. (See Illus. 262.)

*Illus. 262. Using a #48 to make a groove.*

# Plow Planes

Remember, dado planes are used for making grooves across the grain. Plow planes make grooves with the grain. Although most of the combination planes discussed in the next chapter will work as plow planes, many companies made a plane that was used for that one specific purpose.

Some cast-iron plow planes were developed to make grooves for installing weatherstripping in doors. They are called weatherstrip planes. However, they are plow planes and you can use them in your woodworking. Stanley's #248 (made 1936–1943) was intended for use by carpenters when they did weatherstripping. However, the identical #248A was a woodworker's plow plane. The difference between the planes was the number of cutters provided by the manufacturer. The #248 had two cutters (⅛ and ⁵⁄₃₂ inch). The #248A had seven cutters ranging from ⅛ to ⅜ inch. The cutters are very simple, and you can easily make your own.

While a great deal of expense was lavished on some wooden plow planes, turning them into show pieces, metal plow planes were usually straightforward and simple. I own a metal plow plane which, although factory-made, does not even have a maker's stamp on it. Although it works well, it is as basic as is possible to get. (See Illus. 263.)

*Illus. 263. Cast-iron plows are extremely simple, and provide curious contrast with their wooden counterparts. This plane is unmarked. It was perhaps made by a recognized tool manufacturer, but sold under another company's brand. It may have had a paper label that has been lost or removed.*

A metal plow plane is universal and has all the same features as its wooden counterpart. The first is the movable fence that controls the placement of the groove relative to the edge of the board. The fence is usually mounted on two metal rods. Usually, you can move the fence itself along the rods by first loosening two setscrews.

On my plow plane, as on wooden ones, the fence is attached to the ends of the rods. To adjust the fence, you have to move the rods themselves back and forth. On this plane, the rods are secured in the body by two setscrews which can be loosened with a screwdriver. The Record fillister has this same device on it.

A plow also has to be adjustable for depth, and usually has a depth gauge placed ahead of the mouth on the right side. The depth gauge is usually similar to that found on an iron fillister; you can adjust it by loosening a setscrew.

The bodies of most plow planes have a profile similar to that of either a rabbet or dado plane.

They have a closed handle with a blunt toe. These plow planes are usually either nickel-plated or japanned in a dark color.

Some plow planes have a mechanism for longitudinal adjustment. Others, like mine, are adjusted completely by hand. In this case, you free the cutter by loosening two screws with a screwdriver.

## Selecting a Plane

Before purchasing a plow plane, check all the parts and setscrews for damage. Loosen the fence and slide it, to make sure that it moves freely. Verify that all of its parts are still with it. It is common for these planes to be missing irons, but they are so simple you can easily make replacements for any you lack. They are usually made of ¼-inch-thick tool steel, which can be obtained from most industrial suppliers.

Before tuning the plane, disassemble it and clean off any grime. A plow plane has a keel, rather than a sole, so you do not have to joint it. The cutter is supported by the front edge of the rear keel. Iron plow planes are used the same way as are wooden ones (Illus. 264), and their cutters are sharpened the same way. Refer to Chapter 14.

*Illus. 264. Using a metal plow plane. Note that the chip is ejected out of the left-hand side, rather than to the right as on a wooden plane.*

# Circular Planes

During the late 19th century, tool companies developed one of the most ingenious of all special-purpose planes. They called it a circular plane, although it was also known as a compass plane. Like its wooden counterpart, this circular plane is used for smoothing curved surfaces. In writing about the metal side rabbet, I commented that occasionally a metal plane would prove to be far superior to the wooden version. This also applies to the circular plane.

While the wooden plane has a fixed radius and only works a concave curve, the metal plane is adjustable and can be made to fit almost any surface, concave as well as convex. The plane's sole is a thin, flexible strip of steel, usually a little over 2 inches wide at the mouth and a little over 10 inches long. (See Illus. 265 and 266.)

Several tool companies made circular planes. Stanley alone made as many as seven models under its own name, and other models under other trademarks owned by that company (Bailey, Victor, and Defiance). Most of these planes have a short cast-iron body that is mounted near the middle of the sole. On many planes the body and sole are attached by an ingenious device. The body has a dovetail slot cast into its lower surface, running side to side, while the sole has a dovetailed bar riveted to it. (See Illus. 267.) The bar slides into the slot in the casting, attaching it and the sole securely together. The bar and the thin

*Illus. 265. A Stanley #113 circular plane with its sole in a concave position.*

*Illus. 266. A circular plane with its sole in a convex position.*

*Illus. 267. The adjustment gears on the side of a #113. This particular plane dates back to before 1903, when graduated scales were added to the gears. Note the dovetail bar that joins the sole to the casting. Circular planes are still being made today.*

steel sole each have a mouth cut into them through which the cutter projects, and through which the shavings pass into the hollow cavity of the body. (See Illus. 268.)

Circular plane bodies also have one of two arrangements for securing and adjusting the cutter. We have seen examples of both types used on other planes in this chapter. The first is the cap and setscrew, as found on the fillister. This arrangement usually includes a wheel for longitudinal adjustment, but nothing for lateral adjustment. The second arrangement is the modern lever cap, as found on bench planes and the bench rabbet. Planes with a lever cap usually have a lateral adjustment lever and a brass longitudinal adjustment wheel behind the cutter.

Circular planes have some type of handle, often no more than just a curved surface that fits the palm of your hand, between the thumb and index finger. The bodies are japanned in a dark color,

while the other parts are surface-ground and polished.

The most common circular plane you will see on the second-hand market is the Stanley #113. This plane was produced between 1879 and 1942. The one I own was made before 1897. Record still makes its circular planes.

The #113 has some advantages over other models. On some planes, the front and rear of the sole are bowed by separate actions, usually a screw at each end of the plane. The #113 is adjusted by a single large screw mounted on the front of the casting that raises or lowers both the front and back of the sole simultaneously. Turning the screw lifts or lowers a metal bar. The bar is connected to one of a pair of metal gears on the left side of the body. The bar's action is transferred to the rear through these gears. As the front gear moves, it turns the rear gear, which lifts or lowers a similar bar in the back of the plane. On versions

*Illus. 268. The flexible sole is joined to the dovetail bar with two rows of rivets.*

of the #113 made after 1905, these gears are marked with graduations, so that a setting that is used repeatedly (or an exact reverse curve) can be found quickly.

## Selecting a Plane

Circular planes are not rare, but neither are they a common item on the second-hand market. You will probably have to buy one from a dealer who specializes in old tools. Collectors like circular planes, and they cost more than most other tools discussed in this chapter, although they are usually cheaper than a new one. If you buy a plane over the phone or through the mail, ask for a later-model plane, one that has a lever cap, lateral adjustment lever, and graduated gears. These parts are easier to adjust than the older cap-and-setscrew type.

Before purchasing a circular plane, make sure that it is complete. Examine all the parts for damage, paying particular attention to the sole. Over time, continued flexing will fatigue the metal, and it can break. This problem usually occurs near the mouth, where the sole is riveted to the sliding dovetailed bar.

Examine the sole closely under a bright light for tiny fissures that occur before the metal breaks. Do not buy a plane that has this problem, and do not buy one with a broken sole.

Circular planes are intriguing, and anyone who picks one up feels compelled to work its mechanism. To help ensure that your plane outlasts you, discourage visitors to your shop from adjusting your plane unnecessarily.

Before using your circular plane, disassemble it and clean it. You do not have to joint the sole (since it is flexible, you probably couldn't even if you tried). Most of these planes have numerous parts that should be lightly oiled.

## Sharpening the Iron

A circular plane is used for smoothing, not for rough work. These planes do not excavate as do some wooden compass planes. Grind the cutter the same way you would grind the cutter on a smooth plane, using the process described in Chapter 9. The cutting edge should be square to the sides of the iron and either straight or just slightly crested. Test the edge with a try square to see if it is still at a right angle. If not, spray the forward surface with layout fluid and use the square to trace a new edge.

When planing a curve, you will often run into end grain. To ensure that the plane cuts as cleanly as possible, take time to develop as keen an edge as you can. Your circular plane should be as sharp as your best smooth plane. Be sure to flatten the forward surface of the cutter on the lapping table and to polish it on a fine stone.

The cutter is a double iron. When sharpening, also hone the front edge of the chip breaker to help keep chips from catching underneath it.

After you have returned your cutter to the plane, adjust it as you would the cutter on a bench plane. If you have purchased the older cap and setscrew type, there is a longitudinal adjustment wheel either on the right side or in back of the cutter.

If there is no lateral adjustment lever, sight down the sole to determine which side of the cutting edge is high. Loosen the setscrew enough so that the cutter is still held in place but can be moved with your fingers. Shift the upper end of the cutter in the necessary direction and then retighten the setscrew.

# Using a Plane

Set your circular plane to the curve you need to work. When using one of these tools, you have to be constantly aware of the grain's direction. If you are planing an outside curve, you will usually have to start at the apex. In other words, plane downwards. If you plane upwards, you will usually be planing into the grain, which will not result in a smooth surface. (See Illus. 269.)

If you are planing an inside curve, usually you have to start at the top of the contour and plane downward towards the middle. At the bottom of the curve, the grain will usually change direction and you will have to work downward from the other side.

*Illus. 269. Using a circular plane to smooth a convex curve. The tool requires both hands, but I have dropped one to provide an unobstructed view of the tool.*

# 15
# Combination Planes

During the last quarter of the 19th century, tool companies began to develop iron special-purpose planes that were able to perform two or more different functions that had previously been done by separate planes. These were the first combination planes.

A combination plane is different from a universal plane. A universal plane (for example, a plow or fillister) performs only one function, but is adjustable for a range of widths and depths. In contrast, a combination tool performs two or more different functions. To further complicate the issue, combination planes can also be (and often are) universal as well. In such a case, they are able to perform two or more functions, but do these functions across a range of widths and depths.

Although universal and combination planes are different, both were invented for the same reason. A tool that can be used for more than one purpose or in a variety of sizes reduces the number of tools a woodworker needs to own, transport, and store. Anyone who works in a crowded workshop will appreciate this approach to toolmaking, an approach that to this day is prompting the development of the multi-purpose machines that are now being manufactured by a number of different companies.

In 1871, Stanley introduced its first combination plane, the #41, and called it "Miller's Patent metallic plow, fillister, rabbet and matching plane," which is quite a mouthful. Over the next several decades, a tremendous variety of combination planes were developed by Stanley and other tool companies. Besides Miller's Patent, the following are some of the other combination planes produced just by Stanley: a plow and rabbet plane (#54); a rabbet and block plane (#140); a plow and match plane (#44); a dado, fillister, plow, and match plane (#46); a rabbet and fillister plane (#278); a plow and match plane (#43); and a fillister and bullnose (#78), which was described in the last chapter.

During this inventive period, Stanley also developed an adjustable beading plane (#50) that made quirked as well as center beads. Previously, both types of beads had to be made by individual wooden planes that (as explained in Chapter 12) had to be owned in sets. Now, a woodworker could own a single plane that would make quirked as well as center beads in a variety of widths.

In 1884, Stanley introduced the first model of what would become the best-known combination plane made by any company: the Stanley #45. The #45 was produced for 70 years, and during that time improvements and additional functions were added to the original model.

The literature for the plane describes the #45 as "seven planes in one." (See Illus. 270.) Most #45 models make a variety of sizes of beads and various widths of rabbets, dadoes, and grooves, do slitting (cutting strips of thin stock), have a sash cutter, and use groove cutters in combination with a tongue cutter to make matched boards. Later models (made after World War I) also have a variety of hollow and round cutters, and can also make reeds and flutes of varying widths.

Cast-iron combination planes were first developed as an effort to reduce the number of tools a craftsman needed in the shop or at the worksite. However, as these planes became increasingly more complicated and sophisticated, tool manufacturers sought to develop a single tool that could perform all possible functions. Ideally, this would be the only special-purpose plane a craftsman would have to own.

The #45 was a step in the manufacture of an all-purpose plane. The goal was achieved a few years later (1897) with the introduction of the Stanley #55, which the company also called the Universal Combination Plane. The plane was equipped with 52–55 different cutters (depending on the year it was made). It could do all the operations performed by the #45, plus make chamfers and a variety of mouldings. The #55 was advertised as no less than "a planing mill within itself." It was called a "molding, match, sash, beading, reeding, fluting, hollow, round, plow, rabbet and fillister, dado, slitting, and chamfer plane."

While the #45 could make simple moulding profiles (beads, reeds and flutes), the #55 could do all this, and make complex mouldings as well. The plane made mouldings in two different ways. First, it came equipped with a variety of moulding cutters which consisted of several sizes of roman, Grecian and reverse ogees, as well as 9 beads and 2 quarter rounds with beads (cove and beads). Second, using its hollow and round cutters, you could also make an unlimited variety of mouldings in different widths. Remember, wooden hollows and rounds were often used for this same purpose.

Besides the original 55 cutters that were pro-

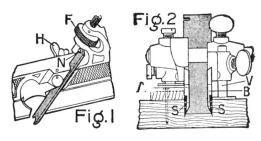

## Directions for Use
*Cuts (except No. 1) show Plane as it looks from the front*

**Fig. 1—CUTTERS**—To insert a cutter, loosen Cutter Bolt (H) and place cutter in position with slot on Pin (N). Adjust by means of Adjusting Nut (F), then tighten Cutter Bolt (H).

**Fig. 2—DADO**—Move Sliding Section (B) up to cutter until its Spur (S) is directly in line with the left edge. Attach extra Gauge (V) to Sliding Section to gauge depth. A batten is used for gauging position of dado.

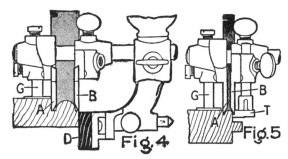

**Fig. 4—BEADING**—For ordinary Beading place Sliding Section (B) so that the outside of same is in line with left side of the cutter. Fence (D) gauges the distance of bead from edge of board, and Gauge (G) depth of cut. Spurs not used.

**Fig. 5—BEADING MATCHED BOARDS**—Attach Beading Gauge (T) to left of Sliding Section (B). This provides a guide above the tongue. Gauge (G) regulates depth of cut. Fence (D) and the Spurs are not used.

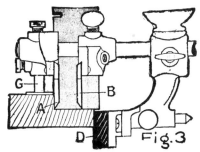

**Fig. 3—RABBET**—Attach Fence (D), putting Arms through upper holes. The Fence regulates the width of the cut and if required, slides under the cutter. The Sliding Section (B) also slides under the cutter, forming a support on the outer edge of the rabbet. Gauge (G) regulates the depth of the cut.

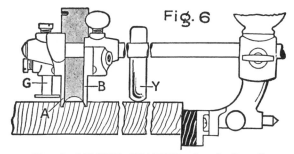

**Fig. 6—CENTER BEADING**—Attach Cam Rest (Y) to either Arm between Sliding Section and Fence to steady Plane. Attach Fence and Plane will cut bead five inches from edge of stock. Extra long arms can be furnished on special order which will permit of a bead being worked eight inches from edge of the board.

*Illus. 270A. These illustrations and those on the following page are from a Stanley pamphlet that shows how to use a Stanley #45 plane.*

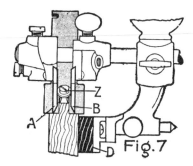

**Fig. 7—MATCHING (Tongue)**—Insert the tonguing cutter and set the Stop (Z) attached to same, at the proper point to obtain the height of the tongue desired. Fence (D) regulates the position of the tongue on the edge of the board. The Gauge (G) and the Spurs are not used. With two cutters, boards varying from three-eights of an inch to one inch in thickness can be matched in the center.

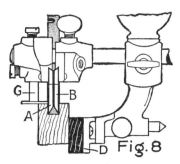

**Fig. 8—MATCHING (Groove)**—Use the one-quarter inch plow cutter. Fence (D) regulates the distance of groove from face of board and Gauge (G) the depth of groove. No Spurs are necessary.

**PLOW**—Use Plane same as in cutting groove for matched boards, except that when cutters are less than one-quarter inches wide, the Sliding Section should be removed. Spurs not used.

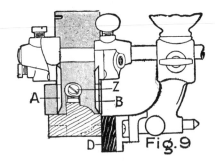

**Fig. 9—SASH PLANE**—The Sash Cutter is similar to the tool used for cutting the tongue on matched boards, as it has a Stop (Z) which can be adjusted to regulate the depth of the cut. One side of the moulding is cut first, the work is then reversed and the operation repeated on the other side. Fence (D) is used as in Matching. Spurs not used.

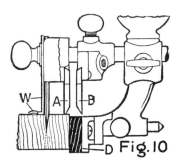

**Fig. 10—SLITTING**—Insert the Slitting Cutter in the slot on the right side of the Main Stock, and just in front of the Handle. Place Depth Gauge (W) over the Blade and fasten both by means of the thumb screw provided. Fence (D) gauges the distance of the cut from the edge of the board. For thick boards slit both sides.

*Illus. 270B.*

vided with the plane, Stanley made an additional set of 41 shapes that could be purchased by mail. In addition, the company sold blank cutters that the woodworker could cut to special shapes. Stanley also advertised that special cutters could be ordered and ground to shapes specified by the customer. Hence, in theory, the #55 was limited only by the user's imagination. It was the woodworking equivalent of the modern food processor.

The #55 seems to have been the only special-purpose plane a woodworker needed. The plane and its numerous parts weighed about 25 pounds and fit into a 10 x 12 x 16-inch box. (See Illus. 271.) The separate planes necessary to do all the same operations a #55 could do would fill several shelves and cost much more.

Several types of combination planes are common on the second-hand tool market. Some of the planes you will find have been discontinued, and some are still being made. Stanley stopped producing most of its combination planes in the early 1960's, but still offers the Stanley Special Combination Plane. Record has been producing combination planes for a long time, and although some of these planes have been discontinued, it still sells its Plough/Combination Plane. Record no longer makes its Multiplane (#405), which closely resembled the Stanley #45, but a similar plane is once again available under the Paragon trademark. Another close copy is being made in India by another company.

As more and more functions were combined

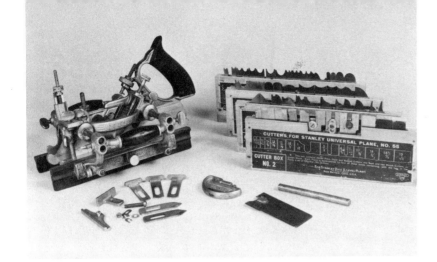

*Illus. 271. A #55 combination plane with all of its parts. (Plane courtesy of Tom Hinckley.)*

into individual combination planes, these tools became more and more complicated. These planes do work and can perform all the functions claimed for them. On the other hand, they do most jobs only satisfactorily, and do not always work as well as the individual planes described in Chapters 12–14.

Combination planes can take a long time to set up for more complicated operations. They have (especially the #55) a large number of parts; these parts have to be specially arranged for each function. You can literally spend more time setting up the tool than you will spend using it.

Only you know the level of performance you need from your tools. If you are a carpenter and have to make occasional rabbets, grooves, dadoes, and simple mouldings (such as beads), you will be very satisfied with a combination plane. If you work wood for pleasure and your income does not depend on the time you spend woodworking, a combination plane might be a good investment. However, if you are a professional cabinetmaker working with figured woods such as mahogany, a combination plane might not work as well as you need.

Perhaps the following information will help you to decide whether or not a combination plane will be suitable in your woodworking. Many combination planes (with the exception of the Stanley #140 and the #78) have a variety of fixtures and devices that regulate depth and width of cut. However, they lack important features that make moulding planes and iron and wood special-purpose planes work well. These features include a sole that conforms to the shape of the cutter, as well as a mouth and throat.

Remember, one of the sole's purposes is to hold down the chip in advance of the cutter. As the chip passes through the mouth, into the throat, the corner formed by the sole and the wear (the front surface of the throat) kinks, or breaks, the chip. If the chip were allowed to lift, it would tear from the surface, instead of being severed.

Many combination planes have a keel or skate very similar to that on a plow. The mouth is the space between the keel's front and rear sections. On a combination plane (such as the #45 and #55), the cutters are little more than straight (or contoured) chisels, suspended in a special harness. They work best on straight-grained woods and, in my experience, are more inclined to tear than moulding and special-purpose planes.

I own a #45, but usually prefer to use individual special-purpose tools, even though they do cost more money (in total) and take up a lot of room in my shop. However, I still find uses for my #45. Sometimes in restoration I need to make a small amount of a simple moulding, but do not have a moulding plane that is a perfect match. I can easily make the exact cutter for my #45. As mentioned in Chapter 12, my #45 can also be used as a V-plow. The cutter can be made out of a piece of old tool steel ⅛ inch thick by ⅜ inch wide, and the cutting edge ground to a V-shape.

If you have decided that a combination plane would be useful in your woodworking, you now have to make another decision. Which one should you buy? I suggest that you begin by thinking about what jobs you want your tool to do: rabbeting, dadoing, grooving, etc. Next, decide which type of plane will fill your needs and refer to some of the books listed in the Bibliography on page 254 to become familiar with that particular plane. Now you will be ready to begin your search.

# The #45 Combination Plane

If I tried to describe every type of combination plane that can be found on the second-hand market, this chapter would be a book unto itself. I will instead describe just the #45. First, it is the plane I own and am familiar with. Second, it incorporates most of the concepts used in other combination planes, including the #55. Third, it (or its close cousin, the discontinued Record #405) is the combination plane most commonly found on the second-hand tool market.

A #45 is made up of three major parts, all made of cast iron and nickel-plated (although some earlier models were japanned black). (See Illus. 272.) The first part is the narrow cast-iron body called the main stock. Standing alone, the main stock looks similar to that of a dado plane. (See Illus. 273 and 274.) However, its tote (which has the same shape as the totes on the dado and fillister) is made of rosewood, which creates a pleasing contrast to the dull, metallic sheen of the nickel plating.

The body does not have a sole, but rather a narrow keel (also called a skate) like that on a wooden plow plane. In fact, its front end is upturned, like the blades on the old ice skates worn by Hans Brinker. Perhaps this upturned end is actually a clever visual pun. The keel is not part of the casting, but is riveted to the main stock. As on wooden and iron plows, there is a break in the skate that acts as the mouth. The cutter rests on the rear section of the skate and is supported by it. This stiffens the blade and helps reduce chatter.

Above the mouth is a depression in the body which (without a true throat) helps to direct the shaving to the left. Also in the casting is a 1/8-inch-wide groove that's angled so that it is a continuation of the rear section's front edge. The cutters are fit into this groove. When a cutter is in place, its right side aligns with the skate so that no lateral adjustment is necessary.

On earlier models of the #45 (made before 1905), the cutters have to be adjusted by hand, the same way they are adjusted on rabbet and dado planes. On later models, the cutter is adjusted longitudinally with an adjustment wheel that is mounted above the cutter and in front of the tote. The wheel is held captive, and when it turns, a screw (threaded through the wheel's center) moves up and down. In the end of the screw is a projecting tooth, which engages a small slot cut into the upper right rear corner of each cutter. When the adjustment wheel is turned, the screw moves the cutter with a slow, precise motion. It is a very easy and exact way to adjust a plane.

On all models of the #45, the cutter is secured in position by the cutter bolt, which is fitted into a hole in the stock just in front of the groove. The bolt has a flat face ground on the side of its head. This face is at an angle so that it can produce a taper. When the wing nut on the right side of the plane is tightened, it draws the angled face against the cutter, holding it securely in place. The bolt's head has a small slot cut in it that engages a locating pin inside the hole. The pin and slot ensure that the bolt is positioned so that the tapered face is always drawn against the cutter.

The stock has two depth gauges mounted on its right side. The front gauge is used to regulate the iron's depth of cut. It is adjacent to the mouth and is adjusted by a captive nut similar to that used to adjust the cutter. It has a lock screw that holds the setting. The second depth gauge is just in front of the tote and is used when the plane is slitting. The gauge has no adjustment screw and is moved by hand. It, too, is held in place by a setscrew.

A spur wheel is recessed into the right side of the skate, just in front of the mouth and behind the front depth gauge. The wheel is identical to those found on the dado plane and the duplex, and is also adjusted the same way. The spur is used when the plane is doing cross-grain operations, such as making dadoes and rabbets.

Two 3/8-inch holes pierce the main stock. One is in front of the tote, while the other is just behind the toe. A round metal rod fits into each hole and is held in place with a lock screw. These rods are the means by which the plane's other two sections are attached to the main stock.

The second major part of the #45 is called the sliding section. It consists of a cast-iron body and another skate, identical in shape to that on the main stock. (See Illus. 275.) This skate also has a spur wheel, but it is recessed into the left side, rather than the right.

*Illus. 272. A Stanley #45 combination plane. This type of #45 was made after 1922, when the name Stanley was added to the sliding section.*

*Illus. 273. The right side of the main stock has a depth stop, a spur wheel, a cutter bolt wingnut, and a slitting cutter stop.*

*Illus. 274. The left (inside) surface of the main stock shows the throat, the cutter bolt, and the cutter adjustment.*

*Illus. 275. The sliding section has another skate like the one on the main stock. It also has a spur wheel, that is mounted on the left side of the sliding section. When setting up the cutter for use, align the sliding section with the cutter's left side.*

179

The body of the sliding section is narrow and has two holes that pierce it so that it can slide on the two rods that connect it to the other two major parts. There are also lock screws above each hole that secure the setting.

The body has a bow in it that curves out to the left. This bow is in line with the mouth and cutter of the main stock. Its shape allows chips to be ejected without interference.

In the front end of the sliding section's casting is a hole for another depth gauge. This one has no adjustment screw; you move it with your fingers and lock it in position with a thumbscrew. When using the plane to cut dadoes, set this depth gauge to the same depth as that on the right side of the main stock.

When a cutter is mounted in the main stock, the sliding section is moved into position so that its skate's outside surface aligns with the cutter's left edge. (See Illus. 276.) In this position, the rear skate of the sliding section supports the cutter in the same way as does the skate on the main stock. When the sliding section is properly aligned, the two spur wheels are also in line with both of the cutter's outside edges.

The third major section of a #45 is the fence. (See Illus. 277.) This part is more complicated than the fences found on plow planes and fillisters. Much of the fence assembly is made of cast iron. This casting consists of two vertical arms connected by a horizontal section.

The #45's fence also slides on the two rods that connect it to the main stock and the sliding section. In this case, the fence's vertical arms have two pairs of holes. These pairs are arranged one over the other, so that by choosing to use

*Illus. 276. You can also adjust the fence, which has a wooden bearing surface, by sliding it along the rods. Note that you can also rotate the fence by turning the adjusting screw (center), a device which was added in 1922.*

*Illus. 277. A view of the underside of a #45 combination plane that is being used as a plow plane. The skates of the main stock and sliding section line up with the corners of the cutter.*

either the upper or lower holes you can raise or lower the fence. When the fence is mounted on the upper holes, its wooden bearing surface is below the cutter. This is important in some operations, such as when using the plane as a fillister to make rabbets. Each pair of holes has one thumb-screw to lock the fence in place.

Mounted over the front pair of holes is a rosewood knob similar to that on the front of a bench plane. The knob helps you control the plane. On earlier models of the #45, this knob was on the front of the main stock. It was moved onto the fence as a later improvement.

The fence's bearing surface is a strip of rosewood. As on plow planes and fillisters, it rides on the edge of the work. This wooden strip can be adjusted horizontally. It too, rides on two short rods that fit in holes located at the bottom of the fence, under each vertical arm. The adjustment screw for the bearing surface is in the center of the casting's horizontal section.

The #45 has other parts that are used in some operations. The plane comes with two pairs of rods, one long and one short. The maximum reach of the fence when mounted on the short rods is 1¾ inches. The long arms are useful when you are grooving or making center beads a greater distance from the edge of a board. Their maximum reach is 5 inches. The plane is less cumbersome with the short rods, and those are the pair I most commonly use.

When the #45 is at maximum settings on the long rods, another piece, called the cam rest, is mounted on the front rod, between the fence and the sliding section. The cam rest is roughly semi-circular, about 2¾ inches in diameter, and has a rounded outer edge. The cam runs on the surface of the wood, and helps to stabilize the plane. (See Illus. 278.)

# Selecting a Plane

Some combination planes (especially the earlier ones, like Miller's Patent) are eagerly sought by collectors, and their prices are high relative to other tools that can perform the same functions. On the other hand, such planes as the #45 and the #55 are readily available, although they usually have to be bought from tool dealers. The prices for these planes seem to be cyclical, rising and falling over a period of time. However, even a combination plane at a bargain is relatively expensive, so do not buy one impulsively.

Carefully examine any combination plane you are considering buying. You can make replacement cutters and some other simple parts. However, do not purchase a combination plane that is missing any of its castings or other special fixtures, as these cannot be easily duplicated.

Most combination planes have many removable pieces. Before buying one, it is a good idea to know in advance how many parts and cutters should come with any particular model.

Most combination planes are expensive. If you buy one through the mail, protect yourself by asking the dealer for a written statement attesting to the fact that the plane is complete. Obtain

*Illus. 278. The #45 mounted on its long rods, with the cam rest attached. The plane is used this way for some operations such as center beading (making a bead on a board's surface rather than its edge).*

a guarantee that the dealer will refund your money if the plane is not as described.

After you have bought a combination plane, disassemble it and clean each part. Most parts are covered by japanning or nickel plating, which helps to protect them from rust. Dirt and grime can usually be removed with mineral spirits. Once the parts have dried, apply a drop of light oil to the adjustment and setscrews.

Some combination planes are complicated tools. Taking the plane apart and putting it back together again will make you more familiar with the tool and help you to better understand how it works. Though it is not unusual to find #45's and #55's that have always been stored in their boxes and are in such good condition that they do not have to be cleaned, still disassemble the tool to make yourself familiar with it and its many parts.

## Sharpening the Iron

There is an advantage and a disadvantage when it comes to sharpening the cutters of a combination plane. Most of the cutters are very simple and are easy to sharpen. However, there are many of them; for example, a complete, later model #45 has 23 cutters, and a complete #55 has an almost unlimited number.

Flatten the cutters you will be using on the lapping table to remove any flaws such as the surface grinding done at the factory. Otherwise,

refer to the other chapters in this book for information on the necessary sharpening procedures for that particular cutter. For example, if you have to sharpen a moulding cutter, refer to Chapter 12.

Combination planes were a product of the late 19th and 20th centuries. The moulding profiles made by a tool's moulding cutters reflect the styles that were popular when it was manufactured. If you are purchasing the plane for reproduction work, the cutters might not produce the exact shapes you need. The proportions of even simple profiles, like a bead, have changed with time. In some cases, you may want to make your own cutters. Buy tool steel from an industrial supplier. Shape the moulding profile using the information in Chapter 12 and anneal, harden, and temper the cutter using the information in Chapter 6.

## Using a Plane

Describing how to use the #45 in all its permutations would require a great deal of space, as each operation requires different attachments and settings. Therefore, refer to Illus. 270, which consists of text and illustrations that have been reproduced from a pamphlet that was published by Stanley in 1955. These illustrations show how the tool should be assembled for each different function. Since the Record #405 was so similar to the #45, I suspect most of these instructions apply to that tool as well.

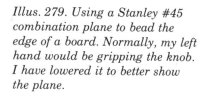
*Illus. 279. Using a Stanley #45 combination plane to bead the edge of a board. Normally, my left hand would be gripping the knob. I have lowered it to better show the plane.*

# 16
# Plane-Related Tools

There are many tools on the second-hand tool market that look like planes, but are not. There are also some tools that do not look like planes even though their functions are quite similar. I have categorized these tools as plane-related tools. Most of these tools have been used by woodworkers for several or more centuries. Through the course of time, some have undergone the same transition planes have: from a wooden body to a cast-iron body. Others have never changed in appearance.

## Spokeshaves

Wooden spokeshaves are one of the most common second-hand tools. I started using them as the result of a conversation I had with a friend. That conversation took place many years ago, back when we were both starting out in our woodworking careers, I as a chairmaker and he as a joiner. I had recently examined a copy of the coat of arms used by the New York City Chairmakers Society during the first quarter of the 19th century. On it was a bit brace crossed over a spokeshave. I understood the dependence chairmakers had on their braces and bits, but did not understand why they would hold a spokeshave in such esteem.

My friend noted that spokeshaves were probably used in many trades, not just chairmaking. "Every woodworker must have used them," he said. "That's the only answer that can explain why they are so common."

He was right. As I continued my research into pre-industrial woodworking, I discovered that spokeshaves were a common tool used by just about every craftsman who worked wood. After purchasing a second-hand spokeshave and teaching myself how to tune and use it, I realized why (in the past) they had been so widely owned. They are wonderful tools; they are able to do many different tasks other tools cannot, and can smooth many areas other tools are unable to reach.

The spokeshave has been used for centuries to clean up curved work or to do light shaping. Although they do not look much like planes, they

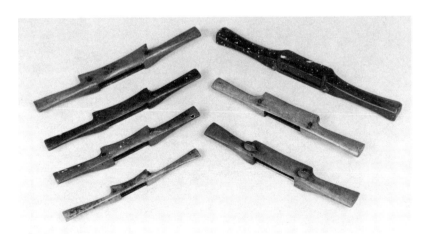

*Illus. 280. A group of spokeshaves. The smallest is slightly more than 9 inches long. The longest (top right) is slightly more than 14 inches. It was handmade. The others were produced by tool factories. They are all beech, except for the smallest, which is made of boxwood.*

are related. As occurred with planes, the earlier spokeshaves had wooden bodies that were eventually replaced by those made of cast iron. However, wooden spokeshaves are far superior to metal ones (my metal one hangs on a hook and has not been used for many years). Since I do not use metal spokeshaves, they fall outside the parameters I set for this book, and are not described in this chapter.

The wooden spokeshave has a long narrow body made up of three separate sections. (See Illus. 280.) The first two are the wing-like ends called the handles. They are oval in cross section, and their widths taper outwards. The third section is the body. This is the center section, and it is long and rectangular. The body's front surface is flat, while the upper surface is usually slightly concave. The lower surface is the most important, as all the working components of the tool are located here.

The front of this lower surface is the tool's sole. Behind the sole is the cutter. The wood under the cutter has been removed to create a throat. The resulting surface slopes away from the sole at about 135 degrees and intersects with the rear edge of the upper surface. (See Illus. 281.) This sloping surface corresponds to the wear on a wooden plane. Remember, the wear is the forward surface of the throat.

The spokeshave's cutter is unusual in shape. (See Illus. 282.) It is a long, thin rectangular blade. It is flat on the bottom and hollow on the upper side, in a manner very much like a straight razor.

The cutter has two tapered vertical tangs, one on each end. These tangs are roughly square in cross section. They are fit into two tapered holes in the body and are held in place by friction. When in place, the cutter covers the throat; its cutting edge is aligned with the back edge of the sole.

During the mid-19th century, another method of holding the iron was developed. Instead of being tapered, the tangs were threaded. In place of the tapered holes in the spokeshave's wooden body are captive nuts in the shape of brass thumbscrews. (See Illus. 283.) The thumbscrew itself has a threaded hole through its vertical axis and turns on a small brass plate (usually diamond-shaped) that is screwed to the top of the spoke-

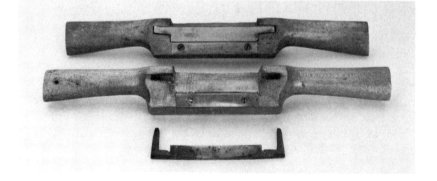

*Illus. 281. Spokeshave soles. The iron has been taken out of the lower spokeshave so that the sloped surface, called the wear, can be seen. Both shaves have been "brassed," re-soled with a strip of brass.*

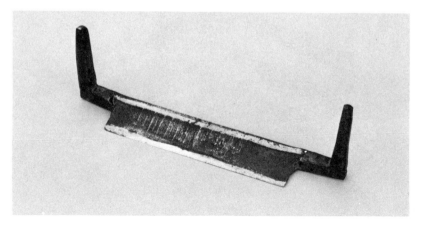

*Illus. 282. The upper surface of a spokeshave's cutter is hollow, somewhat like a straight razor.*

*Illus. 283. A spokeshave with captive nuts. Its cutter has been removed so that the threaded tangs can be shown.*

shave. (See Illus. 284.) Depending on which direction the thumbscrews are turned, they either advance or retract the cutter. These wooden spokeshaves with brass thumbscrews were the deluxe models. Sargent sold this type of spokeshave with a factory-fitted brass strip in the sole.

*Illus. 284. A closeup of a captive nut. Note the threaded hole in the top of the nut.*

No matter whether a wooden spokeshave's iron is held by friction or captive nuts, the blade is set at a very low cutting angle, perhaps only a couple of degrees. Like other low-angled tools (mitre, block, and some shoulder planes), its bezel is placed upwards. Such a placement almost always indicates that a tool is meant to be used on end grain. Indeed, the wooden spokeshave is very useful for this purpose. Its very low cutting angle will pare shavings from end grain rather than create wood dust.

The tool works just as well with the grain. For years, I have used wooden spokeshaves to whittle the spindles for the backs of Windsor chairs.

Most of the wooden spokeshaves on the second-hand tool market were produced in factories.

English companies, such as Marples, were still selling them until very recently. Factory-made spokeshaves are usually made of beechwood, although some are made of boxwood. Like wooden planes, factory spokeshaves all look very much alike.

Some spokeshaves were handmade by individual woodworkers, usually for their own uses. When making one, a woodworker used any wood he chose. The workmanship ranged from excellent to crude. Shapes varied according to the maker's skill and whim.

Wooden spokeshaves come in all sizes, the smallest being very tiny and delicate. These small spokeshaves are usually made of boxwood and are intended for lighter jobs like scroll work, modelmaking, or patternmaking. They also work into a shorter radius than do larger spokeshaves. I own several of different sizes. The longest is more than 14 inches, and the shortest about 9½ inches. Most are slightly more than 11 inches.

Spokeshaves are straight end-to-end, but a variety of curved spokeshaves were produced and you will occasionally come across them in the second-hand market. They come in many different radii, and their wooden bodies, instead of being bent, are cut out of solid wood. Their cutters have the same radii as their soles.

Many of these spokeshaves have handles that are reverse-curved, like sea gull wings. These were used by carriage-makers, and I have not found them particularly useful in general woodworking. I should mention one such tool that I do use, called a travisher. (See Illus. 285.) It is used to finish-smooth the surface of Windsor chair seats which are scooped out, or "saddled," to make them conform to the body. I use my travisher as the final step before sanding.

Spokeshaves were used often, and friction eventually wore a depression in the short sole. In time, the sole was patched with either another piece of

*Illus. 285. The travisher in the foreground was handmade; the one in back of it was made of boxwood by Wm. Marples & Sons, Sheffield, England.*

wood, a piece of bone, ivory, or more commonly a strip of brass. Brassing a spokeshave gave the sole an almost unlimited lifetime, and you will find that many of these tools that were so treated are still very useful in your woodworking.

If you buy a spokeshave that has a worn spot in its wooden sole and you wish to invest the effort, you can also correct the problem by inlaying a brass patch. You can make your own or buy a "mending plate" from a hardware dealer. A mending plate consists of narrow brass strips with screw holes already drilled and countersunk. They are used for mending broken wooden and metal objects. The process is very similar to that described in Chapter 9 for patching a bench plane's mouth. Cut the inlet carefully so that the brass lays flush with the sole. If any correction or flattening is needed use a file or the lapping table.

## Selecting a Spokeshave

Many wooden spokeshaves are simply worn out. Do not buy one of these. When examining a spokeshave, look at the iron to make sure sufficient metal remains. A blacksmith can make a new iron for you, but spokeshaves are so common that I would recommend that you look for one that has plenty of steel.

Check the fit of the tangs in their holes. They should be tight. Look to see if a former owner has filled the holes with shims. Do not waste any time or money on a spokeshave if the tangs do not fit snugly or if the holes have been shimmed. It is very annoying to work with a spokeshave that has a cutter that does not fit securely because it continually loses its setting. This allows the iron to dig into the wood and to chatter and choke.

If the spokeshave has captive nuts, check these for excess play. If they are worn, they too will allow the iron to chatter.

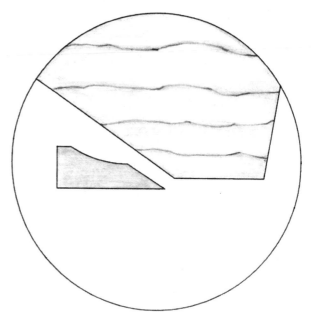

*Illus. 286. A cross section of a spokeshave. The sole on a wooden spokeshave should be flat and parallel to the cutter. Note that the cutter is hollow behind the bezel.*

Occasionally, some previous owner will have reshaped a spokeshave's sole, rather than patch it. You cannot correct this problem without planing away the entire front edge of the shave and gluing on a new piece. Make sure that the surface of the sole on any spokeshave you are considering buying is parallel (or nearly parallel) to the plane of the iron. (See Illus. 286.) If the shape of the sole has been altered by rounding or because this parallel relationship has been changed, the tool will roll forwards when the cutter engages the wood. This will result in a marred surface. It can also pull the iron out of its setting.

A few words should be written about the metal spokeshaves developed in the late 19th century and still sold today. These tools are common on the second-hand tool market. They have so few parts that they are usually found intact. They are practically indestructible and, so, are usually in good condition. Still, I do not use them because wooden spokeshaves are much more superior. After using a well-tuned, properly adjusted wooden spokeshave, most of my students agree with me that metal spokeshaves are not worth bothering with.

Part of the reason why wooden spokeshaves are better to use than metal ones is that the blade on iron spokeshaves is set at 45 degrees, just like the blade on a bench plane. (Some manufacturers made iron spokeshaves that had cutters that were set at a low angle like the cutters on wooden spokeshave. However, these spokeshaves are not common.) This higher angle means that the tool will not cut as cleanly on end grain as the wooden spokeshaves. The higher angle, coupled with a shorter width of sole in front of the mouth, makes the tool more inclined to roll, so that you have to grip it more firmly. This is more tiring on the wrists and forearms.

## Sharpening the Iron

Remove a spokeshave's cutter by tapping the tangs with a light hammer. Make sure you drive them evenly. If the cutter becomes cocked, you can enlarge the holes or even break the tangs. If your spokeshave has captive nuts, you still have to withdraw it evenly.

Once the cutter is loose, you can begin to sharpen it. If the edge is badly damaged, worn, or no longer straight, you will have to grind it. If the steel is not too hard, it is also possible to do this with a file.

The edge is quite thin due to the blade's unique hollow shape. This makes it much easier to overheat on the grindstone, which will draw the temper. The tangs create an additional problem. When you grind the edge, they point towards the grinder and often bump into the motor. This can make it difficult to create a straight edge or to get the correct angle. If you have this problem, use a file.

After you have straightened the edge and created the proper bezel, flatten the bottom surface of the cutter. Be sure to remove any pitting. When you hone, the tangs are again in the way. They are usually far enough apart to fit over the surface of a stone, but may drag against the bench top. Try elevating your stone on a strip of wood. (See Illus. 287.)

Return the cutter to the body and drive the tangs into their holes by tapping their undersides with a hammer. Once again, make sure that each tang advances evenly so that the cutter does not become jammed. If the tangs have captive nuts, turn them at the same rate to make sure that the tangs travel evenly.

When using a spokeshave I elevate one corner slightly more than the other. This means that one end has a heavier set than the other. (See Illus. 288.) If you have to cut a heavy shaving, use the end that is set heavily. Cut a lighter shaving with the opposite side. The center of the iron cuts a medium shaving. Cocking the iron means that the only time you disturb its setting is when you remove it for sharpening.

When you do make adjustments, tap either the top or the bottom of the tang on a metal surface. I usually do this on the jaw of a bench vise. This way I do not have to reach for a hammer.

*Illus. 287. When honing a spokeshave's cutter, elevate the stone on a block of wood so that the tangs do not drag on the bench top.*

*Illus. 288. Before using a spokeshave, I cock the blade so that one end is higher than the other. This way, I have a fine, medium and heavy set all in one tool, and only have to slide the spokeshave until I find the setting I need.*

# Using a Spokeshave

You can use a spokeshave by either pushing or pulling it, depending on the work. If I am working end grain (where there is a greater chance of chatter), I like to have the weight of my shoulders over the tool, and tend to push. (See Illus. 289.) When I am whittling (for example, chair spin-dles), I usually pull the tool towards myself. (See Illus. 290.)

Spokeshaves are so versatile and have so many uses that I cannot list them all. Once you have learned to use one of these tools, you will quickly discover its unlimited potential.

*Illus. 289. Sometimes when using a spokeshave you have to push it. Here a spokeshave is being pushed on the edge of a Windsor chair seat.*

*Illus. 290. Other times you will pull a spokeshave, as, for example, when whittling a Windsor chair spindle.*

# Drawknives

There are two types of edge tools: simple and complex. Thus far, I have only written about complex tools—planes and plane-related items such as the spokeshave. The skill in using a complex tool centers around the ability to tune, sharpen, and set it. Once these steps have been taken, the tool will work nearly the same for anyone, no matter whether the user has a lifetime of experience or is just beginning.

Complex tools all have a sole and a mouth, features that make them adjustable so that you can control the thickness of the chip. However, there is a tradeoff because the user relinquishes the ability to remove large amounts of wood quickly. The chip has to pass through a mouth, so complex tools can only cut a shaving up to that maximum thickness. In other words, the chip you remove can be no thicker than the mouth.

Complex tools such as planes and spokeshaves are intended for finish work or light shaping. However, many woodworking jobs require the rough, fast removal of large amounts of stock. Typically, these jobs are done by simple tools that have open, unregulated blades. For example, hewing (the squaring of round logs into timbers) is done with either an adz or a hewing ax. Both these tools are little more than an open blade that is swung to gain the force necessary to cut thick slices of wood.

A simple tool like a drawknife can be used quickly, but does not have built-in controls; the user provides the control. Unfortunately, the most coarse and basic of tools require the most skill to use. In the hands of a skilled workman, one of these simple tools can be used with remarkable precision. When used by the unskilled, however, these tools can quickly ruin the work. The only simple tools included in this book are chisels, gouges, spoon bits, and drawknives.

A drawknife is no more than a long blade with handles on either end, but is such a useful tool, every woodworker should know how to use one. (See Illus. 291.) Its function is somewhat of a compromise between a hewing tool and a plane. It does have an open, unregulated blade that can quickly remove heavy shavings, but instead of chopping with it, you make slicing cuts while pulling.

*Illus. 291. The upper drawknife has a seven-inch blade and was made by Witherby. The lower one is six inches long and made by the Essex Manufacturing Co.*

Because the knife is unregulated by a sole and mouth, you have remarkable freedom to work very quickly and very close to your final dimensions. In the hands of a woodworker who is skilled, the tool is a precision instrument.

For many years I have taught Windsor chairmaking classes all over the United States. A drawknife is among the tools that I insist each student learns to use. On the first day of class, most students curse their drawknives. By the third day, they bless them. And since my students bring both new and old drawknives to class, I have seen and tried just about every type, and can give you the benefit of my experience.

Drawknives are very common on the second-hand market, and you will come across large numbers and a wide variety of them. Many second-hand drawknives were intended for use in such trades as house building, and are so large that they are cumbersome when used in general woodworking. My knife has a seven-inch cutting edge, and its overall length is a little under 12 inches. Many old knives have the blade length stamped on them. Most also have a maker's mark impressed into the blade.

Many drawknives have a straight blade, while others have a blade with a gentle curve. Both types work equally well. What is more important is that the two handles lie in the same plane as the blade. This gives the drawknife user more control and more efficient use of his muscles. The handles should be as long as the width of the user's palm. Some European knives have egg-shaped handles that are awkward to use.

# Selecting a Drawknife

Before buying a drawknife, make sure that the handles are tight. You have to pull hard on the knife, and cannot work well with one that has handles that pull loose. For safety reasons, the handles should not rotate on the tangs. You do not want seven inches of sharp steel swinging freely.

Before buying, look closely at the blade. Because of the handles, a drawknife is a difficult tool to dress on the lapping table. Do not buy one that is badly pitted, as the pits can take many hours to remove.

Examine the blade's spine. (The spine is the far edge of the drawknife's blade, the edge opposite the cutting edge.) A past owner may have tried to use the knife as a froe (a splitting tool) and may have driven the spine with a hammer. This creates large dents. When these dents drag on the work, they scar the wood. You can remove some small dings with a file, but heavy dents require a lot of work. Sight down the blade to make sure that it has not been bent and that the handles are not out of alignment with each other. (See Illus. 292.)

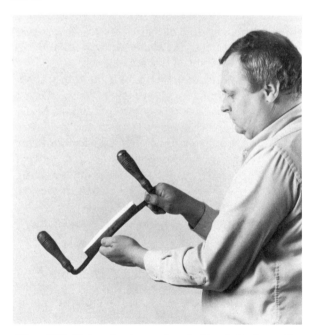

Illus. 292. Before buying a drawknife, sight down the blade to make sure that the handles lie in the same plane and are not out of alignment with each other. They do not necessarily have to be in the same plane as the blade.

Many knives have been ground so often that they have lost much of their width. This results in two problems. One, older drawknives have wrought-iron blades with a layer of forge-welded steel that can eventually be ground away. Secondly, later knives (whose blades are made completely of steel) are hardened only along the cutting edge; the spine and tang are left soft. These areas endure a lot of stress and have to be made tough rather than hard. (See Chapter 6.)

On the market you will often see collapsible knives with handles that are hinged so that they fold over the cutting edge. This seems to make sense, since this type of knife takes up less space in the tool box. The cutting edge is also protected (and equally important, you are protected from the edge). I recommend that you do not buy one of these knives. The handles are too far apart, and although they lie in the same plane as the blade, they are too long. When using one, your hands are too far away from the cutting edge to have good control.

# Sharpening a Drawknife

It is regrettable that so many drawknives are worn out from repeated grindings. This did not have to happen, as drawknives do not usually need to be ground. I have used the same knife for 17 years and have kept it sharp with just whetstones.

Most edge tools work best if the side opposite the bezel is perfectly flat. You do this by lapping that surface on either a stone or the lapping table. The drawknife is an exception. The tool will work better if it has a slight second bezel on the lower surface. This creates what is called a knife edge, which means that both surfaces are slightly rounded. (See Illus. 293.)

I do not purposefully make this second bezel on my knives. Rather, I just leave some of the natural rounding that occurs on that surface due to friction. Periodically, I do some flattening on the lower surface so that the rounding never becomes too exaggerated, but am never so thorough that I remove it altogether.

When you buy a drawknife and sharpen it for the first time, you may have to flatten the lower surface to remove any blemishes that have been caused by rust, wear, abuse, or negligence. (New drawknives are often heavily scratched by the surface grinder and require the same treatment.)

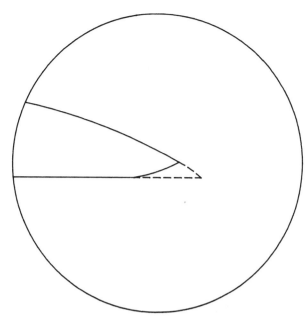

*Illus. 293. A closeup of a drawknife blade. Note how use of the blade has worn a second bezel on the underside. I flatten the bottom of my knife's blade periodically to keep this second bezel from becoming too exaggerated. However, I do not remove it completely.*

A drawknife's two handles make it difficult to use on the lapping table. You will need a lapping table similar to mine (a narrow strip of plate glass); support its underside with a piece of wood. Make sure that the glass is well supported along its entire length so that it does not break under the pressure. (See Illus. 294.)

*Illus. 294. When flattening the bottom of a drawknife on the lapping table, elevate the table so that the handles do not drag on the bench top. Note that the glass is completely supported by a board of about the same width. This prevents it from breaking under the pressure.*

An alternative is to run a stone over the knife. You can rest one handle on the bench top and hold the other in your hand or you can grip one end of the blade in a vise. (See Illus. 295.) In either case, be careful not to cut yourself. This is a job that demands your full attention.

*Illus. 295. When honing a drawknife bezel, I prefer to place one end of the tool in a vise. My fingers are working very close to the sharp edge, and this technique prevents accidents.*

After you have lapped your knife, you can make a second bezel by raising the handles a couple of degrees. You can also make a second bezel with a whetstone. The effect is not a true bezel, but a knife-like edge, one that's slightly convex.

It would be a mistake to overexaggerate the importance of either the angle or the width of this second bezel. When making it, stop and test the edge frequently. The next time you sharpen, use the stone to flatten the back, but do not remove all of the wear that naturally occurs due to friction.

Next, hone the upper bezel. This, too, is a convex surface and is best done with a whetstone. When you are done, you will have created a knife-like edge.

The upper bezel is quite shallow, as low as 20 degrees. Most new drawknives have a very steep bezel. If you buy one of these tools, it will work better if you lower the angle of the bezel with a file.

In time (if you use your drawknife as often as I use mine), you will notice that a hollow spot has begun to wear in the edge. It will occur at the point where you are most inclined to use the knife. (See Illus. 296.) When honing, keep this hollow under control by spending more time working on the higher spots.

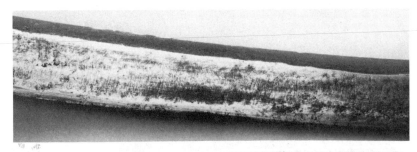

Illus. 296. A depression has been worn in the edge of this drawknife blade at the point where its owner used it most.

# Using a Drawknife

So far, many tools have been discussed that have bezels that have been placed upward; this is necessary so that the tool can be used on end grain. However, drawknives work equally well both with the grain and on end grain.

Many woodworkers hold a drawknife upside down when using it—that is, they ride the bezel on the wood rather than turning the bezel upward. In the hands of the novice, this higher cutting angle reduces the knife's tendency to dive, and gives the beginner a feeling of better control. However, this practice is very limiting, as you cannot take a heavy shaving. You will be able to use your drawknife more efficiently if you hold it correctly and take the time to learn how to control it.

I am not being facetious when I say that you control the knife by making it do what you want, rather than letting it do what it wants. Control is in the grip. Lay the blade flat on the wood's surface. Then raise it slightly to find the angle at which it cuts best. When you have found that position, stiffen your wrists so that when the blade enters the wood it is held rigidly at the cutting angle.

The drawknife cuts best when held askew, as this creates a slicing cut. (See Illus. 297.) When working with the grain, hold the blade at about 20 degrees to the direction of cut. When working on end grain, double that number to about 40 degrees to the direction of cut. Slice with the knife rather than chop. It is not a two-handled hatchet.

When removing large amounts of waste, you can get the blade under the wood and lift the handles. This will split the material loose, rather than cut it. This is an advanced technique and you shouldn't try it until you become more familiar with the tool. Splitting, rather than cutting, is a very fast way to remove stock, but can have disastrous results.

You can cut your finger on a drawknife's open

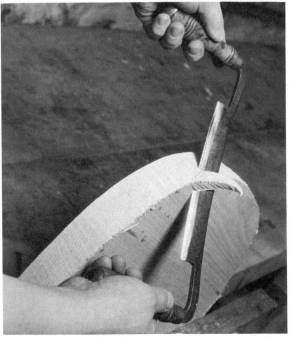

Illus. 297. When using a drawknife, hold it so that its cutting edge is askew. This allows you to cut very thick shavings.

blade if you handle it carelessly. I have used drawknives for 17 years and have never experienced more than an occasional nick.

The safest way to store a drawknife (for you and the tool) is to hang it on a wall. (See Illus. 298.) My drawknives hang behind my bench; each one has its own dedicated space. Two pegs are placed slightly further apart than the length of the blade. The knife hangs horizontally, and cannot swing.

When I travel, I place a wooden guard over the edge of the drawknife. The guard is a piece of 1-inch-thick pine, about 1½ inches longer than the blade. The pine guard has a V groove along its length. I fit the edge into this groove, and then wrap two elastic bands around both the blade and guard to keep them bound tightly together.

*Illus. 298. I keep a guard on my drawknife, even when it is hanging on the wall. The guard that fits over the cutting edge can be made with a plow plane.*

# Wooden Scrapers

Scraping is probably the most ancient technique for smoothing wood. In primitive cultures, it was done with the sharp edge of natural objects such as stone, shell, or bone. In spite of all the planing and sanding machines used by modern woodworkers, they still scrape wood for a final finish because (when done correctly) the process removes shavings that are as fine as tissue. Woods with an erratic grain, like curly and bird's-eye maple, are almost impossible to plane because the grain runs alternately one way and then the other. The only way to smooth them is by scraping or sanding.

The simplest form of scraper is a sheet of steel that is held in the hands. These scrapers are called cabinet scrapers and are very handy for smoothing small areas or for reaching into awkward places. (See Illus. 299.) They have to be slightly bowed, which you can accomplish with your thumbs. For this reason, hand scrapers are very tiring. When used for any amount of time, they also get hot.

Given these drawbacks, it is not surprising that when possible woodworkers prefer a scraper that is mounted in some form of holder. These holders come in many forms that range from the very simple to the very sophisticated. The blades on all these holders are usually held at approximately 70 degrees. In the 70-degree position, the edge cannot get under the chip and lift it. A scraper blade is very light and unable to resist chattering; thus, most scrapers need a means to bow the blade to stiffen it.

The earliest type of body for holding a scraper was made of wood. Most of these bodies appear to have been made by the woodworker for his own use, rather than being manufactured. As a result, there is a great deal of variation in their sizes and

*Illus. 299. The simplest form of scraper is a piece of thin steel with a burr formed on one edge.*

shapes. The most common design of wooden scraper body looks very much like a spokeshave, and dealers very often confuse them. (See Illus. 300.)

You can easily spot the difference between a wooden scraper and a spokeshave by looking at the blade. The blade on a scraper is nearly vertical (the opposite of a spokeshave cutter). The vertical blade results in some other differences between these two tools. Instead of being flush with the sole, the edge of the blade projects very much like that on a plane. Wooden scrapers generally have a throat in front of the cutting edge, through which the shaving passes. The throat on a spokeshave is under and behind the blade.

Most wooden scrapers have a 70-degree inlet cut into the stock. The blade is laid into this inlet. A second piece (with the throat cut into it) is placed over the scraper blade. The sides of the throat create two shoulders that fit against the blade's outer edges. Two wood screws secure this piece to the body and create the pressure that holds the blade in position.

The bodies of most scrapers are made of very hard woods, such as beech or maple. Sometimes,

*Illus. 300. Both of these wooden scrapers were handmade. The one on the left has an ebony sole. Note their similarities to spokeshaves. One difference, however, is that their throats are vertical.*

the sole is made of separate pieces that are glued onto the bottom of the stock. In this case, they are usually made of a very hard exotic wood such as rosewood, lignum vitae, or ebony.

## Selecting a Wooden Scraper

Do not buy a wooden scraper that is imperfect in any way. Wooden scrapers are very common and inexpensive, and most still have a lot of use left in them. I bought the best scraper I could find. It was made by a woodworker who had considerable skills; he took the time to add a rosewood sole.

# Iron Scrapers

During the second half of the 19th century, while wooden planes were being replaced by cast-iron planes manufacturers were also developing cast-iron scrapers. These scrapers all look very much alike. The Record scraper is known as the #080, while Sargent designated its version the #54. Stanley called its product the #80. Stanley's scraper is still in production, as is the nearly identical Kunz brand. This means that you can easily purchase replacement blades if your second-hand scraper needs a new one.

Just as many wooden scrapers resemble wooden spokeshaves, so do cast-iron scrapers look very much like oversized cast-iron spokeshaves. Once again, there are differences that become obvious on closer inspection. The sole of an iron scraper is much longer (front to back) than that on the spokeshave—about 2¾ inches as opposed to 1 inch. This is because the spokeshave has to be able to work in concave curves, while the purpose of the scraper is to smooth flat surfaces.

The body of an iron scraper has a wide, flat bed that rises between the two handles at a 70-degree angle. Across the front of this bed is a flat plate that is attached with two thumbscrews. At the bottom of the bed is the scraper's mouth. (See Illus. 301.) To obtain a setting, pass the blade between the bed and the plate. Then fit it into the

mouth until it projects below the sole. Once you have obtained the setting, secure the blade by tightening the two thumbscrews.

Centered on the rear surface of the bed's lower edge is a third thumbscrew. When this is tightened it pushes against the back of blade, creating the slight bow that is necessary to stiffen it and reduce its tendency to chatter.

Like most other cast-iron tools, the bodies of these scrapers are japanned in a dark color.

*Illus. 301. A #80 Stanley cast-iron scraper. Note the thumbscrew used to bow the cutter.*

## Selecting an Iron Scraper

Like their wooden counterparts, cast-iron scrapers are very common on the second-hand tool market. They are also quite inexpensive. In fact, they often sell for as little as half the price of a new one. You can well afford to be choosy when looking for one of these scrapers. Do not buy one that is less than perfect.

# Scraper Planes

If you do veneering or make furniture that has wide, flat surfaces, such as tables and chests, consider buying a scraper plane (also originally called a veneer scraper). These are easier to use for long periods of time than the cast-iron scrapers described above.

These tools look like bench planes. They have a wide cast-iron body (about 3½ inches) and a knob

*Illus. 302. The sole of an iron scraper. Note that the blade is bowed by the thumbscrew.*

and tote. The sole has a mouth through which the blade projects. (See Illus. 303.)

The mechanism for holding the blade is very different from that on a bench plane. The bed is

*Illus. 303. A scraper plane looks like a bench plane. These tools were made by many companies. The one shown here is a Stanley #112.*

adjustable. It is hinged to the casting, just above the mouth. This allows it to pivot (back to front) from about 88 degrees to about 125 degrees. A threaded rod extends from the back of the bed and passes through a hole in a post. On the threaded rod are two round brass nuts, one on either side of the post. (See Illus. 304.) When you have set the bed at the required angle, screw both nuts against the post. This locks the setting. This device uses the same principal as the two wooden lock nuts on the fence of a wooden plow plane.

The iron is secured against the bed by a screw cap. The cap is not removable, since it is hinged to the bed. The blade is inserted between the cap and the bed and held in place by the setscrew.

*Illus. 304. A scraper plane has a bed that pivots, changing the angle of the scraping blade. To adjust the angle, loosen the two brass nuts and pivot the bed with your fingers.*

This causes the cap to fulcrum on its hinge and press against the blade's lower edge.

## Selecting a Scraper Plane

Scraper planes are not common on the second-hand market, and you will have to buy one from a specialized dealer. The Stanley plane is called the #112. Sargent made two models, numbered #57 and #59. The difference is in their width. The inside of the body on all these planes is japanned, as is the bed and cap. The sole and sides of the body were surface-ground and polished.

The Stanley and Sargent planes have been discontinued. Kunz still makes a very similar plane (also called the #112), so replacement blades are available. The Stanley #80 blade will also fit. Originally, toothing blades were provided with the tool. (See the information on toothing planes presented on pages 200–202.) I do not know if this is still so.

Examine a scraper plane as you would a bench plane. All its parts (except perhaps the blade) should still be with the tool. Its casting should be free of flaws and damage. The knob and tote should be tight, and the hinged bed and cap should move freely. Loosen the screws as much as possible to check this movement.

Before using a scraper plane, first joint the sole on the lapping table.

## Sharpening a Scraper Plane and an Iron Scraper

The blades of wooden and iron scrapers, as well as scraper planes, are sharpened the same way. This technique differs from that used to sharpen a cabinet scraper, even though both rely on a burr to cut a tissue-thin shaving. The blades of these three tools have a bezel, while a cabinet scraper does not.

You have to remove the old burr, which has become dull with use. Burnishing has work-hardened the steel, so the metal will not usually form a satisfactory burr. I remove the old burr with a mill file held in a saw jointer. (See Illus. 305.) This is a cast-iron device used when sharpening saws. The jointer slides over a saw blade, so that the file rides along the tops of the teeth, making all the teeth the same height. It will also slide over a scraper blade. When run on the edge of the scraper, it not only removes the burr, it ensures that the edge is perfectly straight. Saw jointers are sold by some tool catalogues. I bought mine second-hand.

After jointing, regrind the bezel at about 45

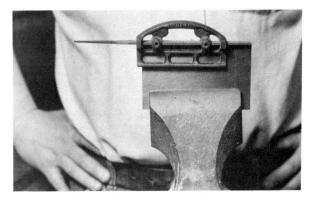

*Illus. 305. An easy way to remove the old burr from a scraper blade is with a 6-inch mill bastard file held in a saw jointer.*

degrees, but do not hollow grind. Hone the bezel on a stone as you would a plane blade, but remember to hold it at 45 degrees rather than at 30 degrees.

Next, flatten the side opposite the bezel on a whetstone. It should not be necessary to use the lapping table unless the blade is rusted or scratched. If it is heavily pitted, purchase a replacement from a tool store or through a new tool catalogue.

Now you can roll the edge to form a burr. You can do this with a hand burnisher, although I prefer a device sold through most of the new tool catalogues, called a wheel burnisher. (See Illus. 306.) This tool has a wooden body with a rabbet cut into the bottom. Both surfaces of the rabbet are faced with metal. A hardened steel wheel is placed at a right angle to the rabbet so that a short section of its perimeter projects into the rabbet's corner. (See Illus. 307.)

Hold the scraper in a vise with the sharpened edge pointing upwards. Put a drop of oil on the edge and spread it along its length. Next, set the wheel burnisher on one end of the edge and pull it over it. The hardened wheel creates a burr with just the right amount of roll.

Return the blade to your scraper or to your scraper plane. Be sure that the burr faces forwards. The set has to be very light. Scrape a piece of scrap wood to determine whether or not the blade needs to be adjusted.

## Using a Scraper Plane

If you are using a scraper plane, you will have to experiment to learn what settings work best on specific types of woods. To begin, pivot the hinged bed forwards to about 70 degrees. Remember, soft woods such as pine do not scrape well.

The scraper plane is a tool used to finish wood and should leave no marks. (See Illus. 308.) Start at one end of the surface you are scraping. The burred edge should be just off the corner of the wood. Make a steady pass and do not stop until the tool has passed over the opposite edge. Should you stop, you will leave a dig in the surface.

## Scratch Stocks

Scratch stocks are simple tools that woodworkers use to make mouldings and other details. They

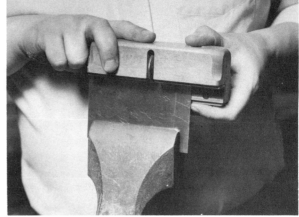

*Illus. 306 (above left). A wheel burnisher has a wooden body with a rabbet cut into it. Projecting into the corner of the rabbet is a hardened steel wheel that rolls a perfect burr every time. Illus. 307 (above right). Using a wheel burnisher.*

*Illus. 308. A scraper plane is a very effective way to finish a wide surface.*

are most useful in hard-to-reach places, such as a curved surface. Scratch stocks are usually made by the craftsman himself for a specific job.

A scratch stock is quick and easy to make. Most have only two parts: a simple L-shaped stock and a blade with the shape of the detail ground into it. (See Illus. 309.) The stock can be cut from a scrap of hardwood, and an old saw or a scraper will provide the steel for the blade. Fit the blade into a vertical slot cut in the stock and position the detail in the corner of the L shape. A screw or a bolt is sufficient to pinch the stock so that it holds the blade in place.

The blade is vertical and cuts by scraping. This allows the scratch stock to work in both directions: with the grain and against it. Because scratch stocks scrape, they work best on hard woods.

The long leg of the L-shape body is the handle. The short leg is the fence, which in use is placed against the edge of the wood. When the scratch

stock is drawn along the edge, the blade scrapes out the shape of the moulding or other detail. (See Illus. 310.)

You will find old scratch stocks on the second-hand market, but remember, they were made for very specific purposes. Unless you have that same exact need, the tool is little more than a curiosity, and is better off left for a collector.

During the 19th century, Stanley produced a tool that was a sophisticated scratch stock. It was adjustable and came with a variety of different blades. The company designated it the #66 and also called it the Universal Beader. (See Illus. 311.) I do not know if any other companies made a similar tool.

The #66's cast-iron body looks very much like that of an iron spokeshave, but is about the same size as the iron scraper. Early models were japanned in black, but most of the ones I have seen are nickel-plated.

Like a scraper, the #66 has two curved handles

Illus. 309 (above left). A scratch stock that's used to make a ⅜ inch bead on curved edges. Illus. 310 (above right). Using a scratch stock to bead a convex curve.

Illus. 311. The #66 Universal Beader is a sophisticated scratch stock used for making simple mouldings and grooves on curved surfaces. To move the fence, loosen the thumbscrew near the right handle.

Illus. 312. The Universal Beader has a movable fence that slides side to side in a slot cut into the sole. The mouth is the opening just below the thumbscrew. This cutter shown here makes an ⅛-inch groove.

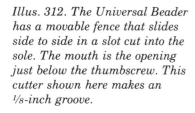

with a bed between them. The bed is also angled forward at 75 degrees to the sole. In the center of the bed's front surface is a shallow, vertical inlet, ⅝ inches wide. A cutter is set into the inlet and then held in place by a special C-shaped clamp that is tightened by a setscrew.

The mouth is directly below the inlet. (See Illus. 312.) The cutter's shaped end fits through the mouth and projects below the flat sole. Another slot is cut into the sole in front of the mouth; it runs side to side in the same direction as the handles. The slot is for the tool's adjustable fences. A thumbscrew fits through the slot and is threaded into the narrow and rectangular fence. To adjust the fence, loosen the screw and slide it to the position you desire.

As on a plow plane, the fences on a scratch stock regulate the distance between the cut and the edge of the wood. The #66 has two different fences that can be interchanged. The first has straight sides and is used when working along a straight edge or vertical curve. The second fence has curved sides that allow the tool to be used on a horizontal curve.

The #66 was provided with six cutters. Each has two different profiles (one on each end) for a total of 12 shapes. As the name Universal Beader implies, most of these cutters make different-sized beads. Some made single beads. Others made beads in clusters of two and three. (When beads are in clusters, they are called reeds or reeding.) One of the other six cutters makes flutes in two different widths, while another cutter makes two sizes of grooves (¼ and ⅛ inch). The cutters are very simple, and you can easily make additional ones that suit your particular needs.

A #66 does not require much tuning. Its sole is flat, but does not need to be jointed. If the casting is dirty, mineral spirits will remove a lot of the grime.

# Selecting a Universal Beader

A universal beader is a simple tool, although not as simple as a scraper. Do not buy one that is damaged or that is missing any pieces except for cutters. These are so simple that you can make your own.

# Sharpening the Iron

To make a cutter, you will need a strip of steel ⅝ inch wide by 2¼ inches long. The strip should be ¹⁄₁₆th inch thick. Use files and/or a grinder to make the desired shape. It may help you to first spray the surface with layout fluid and trace the profile.

The cutter does not have a bezel. A scratch stock is a scratch tool and cuts by scraping. Therefore, your cutter has the same shape on both sides.

The scraping is not done by a burr, but rather by the arris (the sharp corner) of the steel. As the tool is used, the arris is dulled by friction and becomes rounded so that it can no longer scrape a shaving. (See Illus. 313.) When the tool begins to create heat and burnishes the wood rather than cutting it, you have to stop and resharpen the cutter.

Sharpening (as opposed to making a new cutter) is done with files. Use appropriate shapes, depending on the details. (See Chapter 8.) Chain-saw files, for example, come in different diameters and are handy for different-sized beads. Needle files are small and come in a variety of shapes. Run the file directly across the end of the steel at a perfect right angle to the cutter's surface. You do not have to remove a lot of metal, only recreate the sharp corners. (See Illus. 314.)

The groove cutter is the exception. The ends of the blade are bent forwards so that they will cut rather than scrape. Groove cutters have a bezel of approximately 20 degrees ground into them, which is enough to give them sufficient clearance. When you sharpen, lay the bezel on a whetstone and hone it. The upper surface is very short, but if you are careful, you can polish it on the edge of a slipstone.

Return the freshly sharpened cutter to the tool. Place it in the inlet and slide it under the special clamp. Lower the profile through the mouth and adjust its depth. Then, tighten the clamp to secure the setting.

The proper adjustment depends on what you want to do. If you are making a bead, make sure that the entire profile is visible below the sole. The lowest point on the profile (this corresponds to the top of the finished bead) should be almost flush with the sole.

# Using a Universal Beader

When you use the #66, select the cutter profile you want and place it in the plane. Next, select one of the two fences. This depends on the shape of the piece of wood you are working. Adjust the fence so that the moulding shape is positioned where you want it to be in relation to the edge of the wood.

The #66 works best on hard wood, although the groove cutter does well on pine. Place the fence against the edge of the wood and draw the beader towards you. Because the bed is set forward at 75 degrees, the #66 only cuts in that one direction. The #66 works like a moulding plane in that each pass should create more of the profile. When the sole runs on the surface of the wood, it acts as a stop. At this point (if you have set the cutter to the proper depth), the profile should be complete. (See Illus. 315.)

The fence has a groove cut in it so that it can be positioned to cover all or part of the cutter. If the size bead you want to use is in a cluster, cover all but one. If you make your own cutters so that they have shapes other than just beads, the ability to position the fence to use all or just part of a detail gives you a great deal of flexibility.

If you leave a cutter in the beader when you are not using it, slide the fence to cover the exposed end. This protects it from damage and protects you from an accidental cut.

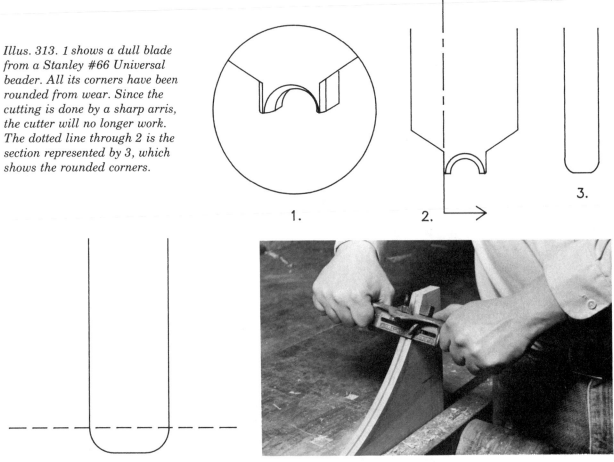

*Illus. 313. 1 shows a dull blade from a Stanley #66 Universal beader. All its corners have been rounded from wear. Since the cutting is done by a sharp arris, the cutter will no longer work. The dotted line through 2 is the section represented by 3, which shows the rounded corners.*

1.

2.

3.

*Illus. 314 (above left). This is a larger view of 3 shown in Illus. 313. Using needle files (and perhaps chainsaw files), file directly across the cutter at a right angle to the surfaces. This will restore a sharp, crisp arris to all corners. You do not need to remove a lot of metal. File only to the dotted line shown here. Illus. 315 (above right). Using a Universal Beader to make a groove along a concave curve.*

# Toothing Planes

The toothing plane is another special-purpose plane that has a vertical blade. The difference between this tool and a scraper is the toothing plane's unique cutter. It has a serrated edge, which looks like a fine saw. (See Illus. 316.) The serrations are created by corrugations in the blade's forward surface. The corrugations are made in the same manner as on a single-cut file, only the cuts are perfectly vertical to the edge, rather than at an angle. When a bezel is ground on the toothing plane's iron, serrations automatically occur on the cutting edge.(See Illus. 317.)

In the days when veneering was done with hot glue, cabinetmakers commonly used the toothing plane to roughen both the back of the veneer and the groundwork to which it would be applied. They thought that the roughness produced a better bond. Although this is no longer a common practice, the toothing plane still has some other uses that will be described below.

Most of the toothing planes on the second-hand market have the same coffin shapes wooden smooth planes have. They also have the same shaped wedges. The difference is the bed, which is vertical (or on some planes, nearly vertical), rather than set at 45 degrees. Also, the iron does not have a chip breaker.

*Illus. 316. A toothing plane. It has only an owner's stamp.*

*Illus. 317. The forward surface of a toothing plane cutter is corrugated. A bezel has been ground on the other side; this has resulted in a series of teeth that scuff a wooden surface evenly.*

Toothing planes were made in the same English and American factories that produced bench and moulding planes. Therefore, they share much in common with these tools. You will usually find a maker's mark on the iron, and the maker and previous owners regularly stamped their names into the heel or toe. The plane's body and wedge are usually made of beech, although other native hard woods such as maple and birch were also used.

Over the years, the average toothing plane did not usually receive the same amount of wear as did the average smoothing plane. Therefore, most toothing planes on the second-hand market are still in good condition. They will not require much tuning to make them useful, and given their function it is not usually necessary to joint the sole.

# Selecting a Plane

Toothing planes are less common than smoothing planes and more expensive. Be very selective when buying one of these planes and only accept one that is in perfect condition.

There are only three parts: the body, the wedge, and the cutter. Examine the body as you would that on a smooth plane. Check the cutter to be sure that it is not heavily rusted. A thick buildup of rust can damage, and even eliminate, much of the corrugation that produces the serrated edge.

# Sharpening the Iron

Since the toothing plane is used on highly figured woods, it is important that the serrations be sharp. Test them with your fingertips. They should catch on your skin. Another way to determine if they have become dull is to hold the iron so that a bright light shines directly onto the ends of the teeth. Light will glint from the tips if they have become rounded from wear.

While examining the teeth for sharpness, test the edge to be sure that it is square with the sides of the cutter. If it is not, you will have to make it square by grinding the cutter on a grinding wheel. Otherwise, you do not have to grind a toothing plane iron very often. The serrations can be restored on a medium-grade honing stone. However, *do not* lap the corrugated surface, as this will wear away the corrugations that create the teeth.

After sharpening, put the iron back into the throat. Keep it in position by putting your fingers over the mouth and allowing the teeth to rest on your fingertips. Insert the wedge and press it so that it now keeps the cutter from moving. Using a light hammer, make your lateral adjustment by tapping on the sides of the cutter. (See Illus. 318.) Lateral adjustment is critical. The row of teeth have to be perfectly parallel with the sole. If one side of the cutter is lower than the other, the scratches will not all be of a uniform depth. You will have to do much more scraping to remove the deepest ones.

When the tips of the teeth are parallel to the sole, make your longitudinal adjustments. Do this in the same manner you would to adjust the cutter on a smooth plane. Tap the heel to retract the cutter, and tap the upper end to advance it.

Illus. 318. Even though a toothing plane has a vertical iron, it is adjusted just like a wooden bench plane. Make sure that the serrated edge is square with the sides of the cutter and parallel to the sole.

The longitudinal setting has to be light. The teeth must project only slightly below the plane's sole so that each individual tooth produces its own separate scratch. Remember, you are only roughing the wood, not cutting a shaving. If the set is too heavy, the teeth will dig into the surface of the wood. The plane will be very difficult to push, and if you do manage to push it you may damage the surface.

# Using a Plane

In the days before the belt sander, a cabinetmaker also used a toothing plane when smoothing such heavily figured woods as curly and bird's-eye maple. I still use my toothing plane for this purpose. First I use it to create a uniform roughness on the board's surface. This is done by pushing the plane in every direction: with the grain, at an angle to the grain, and directly across the grain. (See Illus. 319.) Then, I scrape the roughness until the surface is smooth, leaving no tears in the surface. (See Illus. 320.)

I also use my toothing plane when veneering, but not to rough the back surface of the veneer. I like to reproduce American Federal furniture (1790–1815) in the styles that were popular in Boston and the North Shore (a string of coastal cities north of Boston). Cabinetmakers who worked in these cities were very fond of the flame or feather pattern obtained by cutting veneers from the crotch of the white birch tree (*Betula papyrifera*).

Illus. 319. The toothing plane shown here has just been used to scuff an area on this curly maple board. The scuffing has created a fine dust.

Illus. 320. A scraper removes the scuff marks and leaves a surface that is as smooth as glass.

White birch is a very common wood in New Hampshire, and so I am able to use my firewood as the raw material to make my own veneers. My cordwood dealer sets crotches aside for me, which I resaw into thin sheets. After the veneers have dried and have been laid, they are not smooth. The shrinking that occurs with drying has resulted in some areas that are higher than others.

The grain in the flame pattern is so convoluted that it cannot be planed. Instead, I use my toothing plane to rough the veneer by working the tool in every direction in the same manner used on curly maple. This quickly trims the high spots and creates a flat surface. Then I scrape the veneer with a scraping plane (once called a "veneer scraper") to produce a finished surface. The birch crotch could be smoothed by just scraping, but that would take more time, as that process removes such a fine shaving.

# Miscellaneous Tools

# 17
# Chisels and Gouges

In the section on drawknives in the last chapter, I explained the difference between simple and complex tools. Because they have such features as a sole and a mouth, complex tools can be adjusted for depth of cut. Simple tools are usually an open blade with some form of handle.

Chisels are simple tools. They are little more than long, flat blades, sharpened on one end, with a handle on the other. Gouges are similar to chisels, except that their cross sections are curves.

There are a large number and a great variety of chisels on the second-hand tool market. There are fewer gouges, but they are still very common. As is true of any other tool that is available in large supply, chisels and gouges are quite inexpensive.

As with moulding planes, chisels are often sold "by the inch." Thus, a larger chisel is generally more expensive than a smaller one. Other factors such as quality and condition are ignored. Those tools in excellent condition are often priced the same as those that have been badly abused.

There are many types of specialty chisels—for example, those used for carving and turning. There are also some single-purpose chisels: for example, the swan's-neck is used for making lock mortises, and corner chisels are used for squaring very large mortises. I did not include these chisels here. This chapter is limited to those chisels and gouges that are generally used in cabinetmaking and joinery and are considered bench chisels. I have also included mortise chisels.

All chisels and gouges have two major parts: a blade and a handle. The shapes of these parts are a function of the tool's purpose. Some chisels and gouges are used for heavy work, and are often driven with a mallet. Their blades and handles will be stout. Others do delicate work, such as trimming joints. Their blades and handles will be more delicate.

Chisel and gouge handles have traditionally been made of wood, and this material is still used today. During the middle part of this century, some companies began to use plastic for the handles of some inexpensive chisels that were usually of the handyman quality sold at hardware dealers. However, recently other tool companies started producing quality chisels with high-impact plastic handles. These are usually sold through new tool catalogues. Too little time has passed for these chisels to become a common item on the second-hand market, and you will probably not come across them.

Chisel and gouge handles are attached to the blade in one of two ways: either with a tang or a socket. The socket on a socket-type chisel is shaped very much like a small ice cream cone. (See Illus. 321.) The blade-end of the handle is reverse-tapered so that it fits into the socket. The angle of taper creates what is called a locking taper. The friction in a locking taper causes the handle and the socket to become firmly attached, and they cannot be easily separated. That is why the handle of a socket-type chisel does not readily pull loose from the blade.

As a woodworker, you know that wood handles experience seasonal movement (the wood shrinks and swells, depending upon its moisture content). If you leave a socket chisel untouched through several summers and winters, the wood's seasonal movement might loosen the locking taper. However, if you turn the chisel blade so that it is directed upwards and tap the handle on the bench, you will re-establish the lock. The advantage of a socket is that the handle can be easily removed. A new one can be quickly made in a lathe should the original be broken.

A socket chisel is more expensive for a tool company to manufacture than one with a tang. As

*Illus. 321. The socket on a socket-type chisel.*

*Illus. 322. An exploded view of a chisel with a tang.*

a result, socket chisels always were (and new ones still are) more expensive. However, this is not necessarily the case on the second-hand market, where the relative value of different types of chisels has not yet been established. You will often find socket chisels priced the same as those with a tang.

A tang is a pointed metal shaft that projects from the back of the chisel blade; the handle is mounted onto it. A hole is first drilled into the handle, and then the tang is fit into the hole. At the base of the tang is a small flange called a bolster. The bolster prevents the handle from driving forward on the tang when the tool is used. Without the bolster, the tang would act as a wedge and split the handle. (See Illus. 322.)

There is a metal hoop called a ferrule located just above the bolster on the handle of any chisel with a tang. The ferrule is further insurance against splitting. This metal ring is fitted over the blade end of the handle before it is mounted on the tang. Ferrules are usually made of brass, but iron was also commonly used.

The handles of some tang-type chisels, especially those that are meant to be driven with a mallet, are provided with even further protection against splitting. They have a leather washer between the handle and the bolster that acts as a cushion.

Two other features also help prevent damage to a wooden handle. These features are sometimes found on both socket- and tang-type chisels and gouges. The first is a leather tip on the drive end of the handle—the end that makes contact with the mallet. The leather cushions the blow. A handle with a leather tip is said to be "leather headed."

Other handles have a metal hoop at the drive end that prevents the wood from mushrooming and splitting. This ring is usually made of iron, and the handle is "hooped." A chisel or gouge with

a ferrule (tang-type) that also has a hoop to protect the upper end of the handle is said to be "double hooped." (See Illus. 323.)

Most edge-tool makers produced a complete selection of chisels with all of the features just described. However, they also sold many tang-type chisels without handles. As a result, their prices were considerably lower. The woodworker who bought such a chisel would presumably either make his own handle or buy one, thus saving some money. The companies that made chisels also sold a complete range of both tang and socket-type handles.

The length and weight of chisel blades vary according to the tool's function. Also, some chisel blades have bevelled edges. The bevels are on the same side as the bezel. It took the manufacturer more time to make chisels with bevelled edges and, as a result, these tools (when new) were more expensive than those with square edges. This does not necessarily mean that bevelled-edge chisels found on the second-hand market are more expensive than other chisels.

# Types of Chisels

## Firmer Chisels

A chisel that is struck with a mallet endures a lot of shock. Therefore, the blade on this type of chisel is generally shorter and stouter. These chisels are called firmer chisels (possibly a corruption of the word former or forming). Firmer chisels are general-purpose tools, made to do heavy work at the bench or on the jobsite.

The blades of firmer chisels, as measured from the cutting edge to the shoulder (the corner of the blade just in front of the socket or bolster), range from 5½ to 8 inches long and usually from ⅛ inch to 2 inches wide. (See Illus. 325.) They are nor-

*Illus. 323. Two chisels and a gouge. The upper two tools have tangs. The first chisel is double hooped and has bevelled edges. The lower chisel has a socket and a leather head.*

*Illus. 324. The bolsters and tangs of tang-type chisels. The top chisel is a ⁷⁄₁₆-inch mortise chisel made by A. Mathieson & Son, Glasgow, Scotland. The lower is a ⁷⁄₈-inch bevelled paring chisel made by Buck Bros., Millbury, Massachusetts.*

*Illus. 325. These firmer chisels range from ¼ to 1¼ inches wide.*

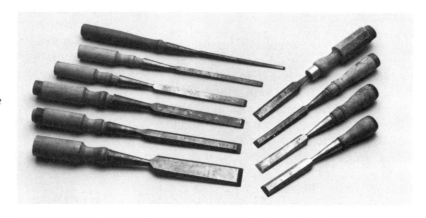

*Illus. 326. These butt firmers range from ½ to 2 inches wide.*

mally graduated by ⅛ths of an inch up to 1 inch: for example, ⅛, ¼, ⅜ inch, etc. From one inch on they either continue to be graduated by ⅛ths or by ¼ths (1¼, 1½ inch, etc.) They can be either the socket or the tang type of chisel. Some firmers have bevelled edges, but most do not.

Firmer chisels also include the shorter butt chisel (also called a butt firmer). The blades of butt chisels are usually about 3 inches long. They also range from ⅛ inch to 2 inches wide, and are graduated the same way as firmers. (See Illus. 326.)

# Paring Chisels

Paring chisels were made specifically for doing fine work such as trimming or fitting joints. They are more commonly used by cabinetmakers, patternmakers and joiners. These tools are much lighter than other chisels and are not intended to be driven with a mallet. They were generally pushed, or at most driven, with the palm. Paring chisels are usually longer than firmer chisels (paring chisel blades average approximately 9 inches in length) and usually range from ¼ to 2 inches wide. (See Illus. 327.)

Most paring chisels have bevelled edges. These edges not only look more attractive, they make it possible for the chisel to work in a corner that is less than 90 degrees, such as a dovetail.

### SLICKS

A slick is a very large paring chisel, and is the largest member of the chisel family. A slick can measure up to 30 inches long if you include the handle, and can have a cutting edge up to 3½ inches wide. (See Illus. 328.) Slicks are usually socketed. The blade may be bevelled, but not always in the usual way. The blade has a spine that runs up the center of the upper surface. Each side of the blade tapers to a thin edge. The cross section is a very low triangle. On some slicks, the cross section is an arc.

A slick is a paring chisel and is usually pushed from the shoulder. A long handle with a knob on the shoulder end makes it easier to push the slick. Slicks are made for very heavy work such as house framing or boatbuilding, and are not usually used at the bench. I have mentioned them here because they are common on the second-hand market; all of the general tool information (as well as sharpening and buying guidelines) applies to them.

# Mortise Chisels

Mortise chisels are another type of chisel used at the bench that can be commonly found on the second-hand market. (See Illus. 329.) As their name implies, these chisels are used for making mortises. This is done by driving them vertically, usually with forceful blows from a mallet. This action chops the wood grain in the mortise into short sections, very much like square plugs, that are then levered loose.

In order to withstand this driving, as well as the pulling necessary to break loose the waste, the blade must be more substantial than even the blade on a firmer chisel. In fact, the tool's most obvious characteristic is its blade's heavy cross section. It is at least square, and usually rectangular. This means that it is thicker than it is wide.

*Illus. 327. A ⅞-inch paring chisel, made by Buck Bros., Millbury, Massachusetts. Note how thin the blade is, the better to reach into confined areas.*

*Illus. 328. A slick. A ⅝-inch firmer chisel provides a sense of scale.*

*Illus. 329. These mortise chisels range from ⅛ inch (left) to ¾ inch wide. Some have tangs, and others have sockets. One is hooped, while another is double hooped. The double-hooped chisel, on the far right, was handmade by a blacksmith. The others are factory-made.*

English and American mortise chisels are graduated by 16ths of an inch. One-eighth inch is the smallest common size, and it is unusual to find one larger than ¾ inch.

Handles on larger mortise chisels are often oval. On smaller chisels they are usually round. Handles are attached with either a tang or socket. The blades are not bevelled.

## Types of Gouges

Like chisels, there are firmer and paring gouges. The length and thickness of the blade are the factors used to classify a gouge as either a firmer or a paring one. There are also socket and tang-type gouges. Gouges do not have bevelled edges. (See Illus. 330.)

Gouges also differ from chisels in that their blades are curved from side to side. The amount of curve is called the sweep. There are three different sweeps. From the flattest to the most curved, they are: the "shallow," "middle," and "regular" sweep. (See Illus. 331 and 332.) You can easily recognize the sweep by looking directly at the cutting edge.

The bezel of a chisel or a gouge is called a cannel. So far I have referred to it as a bezel to maintain consistency. Now that gouges are being discussed, however, the term has to be clarified. The cannel of a gouge can be ground on either the inside surface of the curved blade or the outside. When it is ground inside, it is called an in-cannel, and an out-cannel when ground outside.

An out-cannel gouge is used for excavating—in other words, making a depression in a flat surface. An out-cannel gouge can also be used to make plunge cuts. When a plunge cut is being made, the tool is held vertically. An out-cannel gouge will only make a convex cut. If it is used to make a concave cut, the angle of the bezel will force the gouge forward. (See Illus. 333.) This will result in a sloped, rather than a vertical, cut. For the same reason, when you make a plunge cut with a paring chisel turn the bezel away from the wood.

Concave plunge cuts are made with a gouge that is ground in-cannel. Both concave and convex plunge cuts are used in such jobs as coping mouldings so that they butt tightly against each other or terminate neatly. (See Illus. 334.)

Most gouges are ground out-cannel. If you need one that is in-cannel, buy a gouge of the desired size and alter it. (See the section on sharpening.)

## Selecting and Reconditioning a Chisel or Gouge

There are large numbers of chisels and gouges on the second-hand market. Unfortunately, most have been abused, and many are not worth buying. They have been driven with hammers, so their handles are either broken or missing. Socket chisels have been used after their handles have broken apart and hammer blows have distorted the top of the socket. Old chisels have been used as levers and pinch bars, and in time have become bent, twisted, or broken.

Even though many chisels have been damaged, you will find plenty that are still in good condition. These are often worth purchasing and using.

Before buying a chisel or gouge, examine the steel on the side opposite the bezel. Surface rust can be removed, but the steel should not be pitted near the cutting edge. Pitting anywhere else on

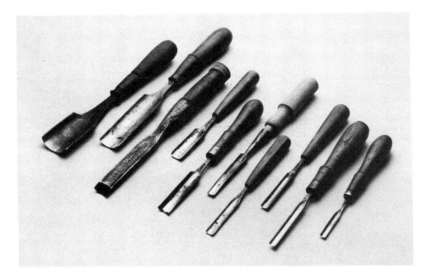

*Illus. 330. A group of gouges. The third and fifth gouges from the left are ground in-cannel.*

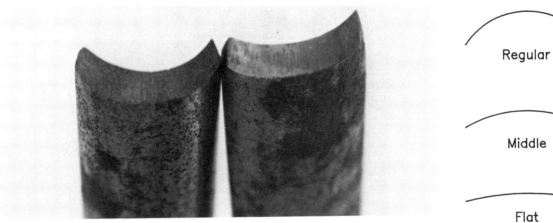

Regular

Middle

Flat

*Illus. 331 (above left). Two ½-inch gouges. The one on the left has a regular sweep. The one on the right has a more shallow middle sweep. Illus. 332 (above right). Three different gouge sweeps. They are, from the top, the regular, the middle, and the flat sweep.*

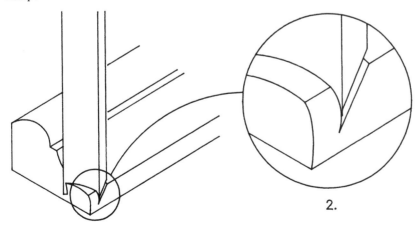

1.

2.

*Illus. 333. It is very difficult to make a concave plunge cut with an out-cannel gouge (1), as the edge is forced forward by the angle of its bezel (2).*

Illus. 334. Coping is an easy and clean way to end a moulding so that it will butt cleanly against another moulding. It consists of carving the profile across the end grain at a right angle to the moulding. The moulding shown here is an astragal and cove, commonly used as a chair rail. The concave cove is coped with plunge cuts made with an in-cannel gouge. A piece of waste wood protects the bench top.

Illus. 335. Cope a convex astragal with an out-cannel gouge.

Illus. 336. The marks in the waste show how this moulding was coped.

the blade only affects the tool's appearance, not its function.

Examine the blade for any fissures or cracks, as these cannot be removed. If they occur near the edge, the tool cannot be sharpened properly. If a crack occurs elsewhere, the blade might break while the tool is being used.

Sight down the blade to determine if it is bent or twisted. Do not be concerned with damage to the bezel or cutting edge, as you can take care of this when sharpening the blade.

Chisels are so common that it is seldom worthwhile to restore one. However, you may decide to buy one that needs to be restored because it is a very high-quality tool, one of a set, or has sentimental value.

If the blade is bent or twisted, do not try to correct the problem without heating it, as it might break. Take the tool to a blacksmith, who can heat the blade in a forge and straighten it on an anvil. The piece will then have to be heat-treated. If the blacksmith does not know how to heat-treat the tool, follow the directions in Chapter 6.

A broken or a missing handle is a very common problem with second-hand chisels and is one repair that you, as a woodworker, can easily do. First remove any part of the handle that still remains. Splitting away the old wood with another chisel is a quick method. Grip the blade of the chisel you are working in a vise. After you have split away one or two pieces of the old handle, it should pull loose.

The handles on socket-type chisels and gouges are often broken off at the top of the socket, leaving the taper securely locked inside. Do not try to burn the wood out with a torch. By the time you are finished, the steel will have become so hot that it has lost its temper. I have successfully removed these nubs by drilling a small hole in the exposed end, screwing a wood screw into the hole, and gripping the screw's head in the claw of a carpenter's hammer. Pull straight back. If this does not work, drill as many holes as possible in the nub. This will weaken the wood so much that pieces can be broken loose and picked out. After enough pieces have been removed, the friction that creates the locking action will be broken.

Now you are ready to mount a new handle. Many hardware dealers still sell them. You can also make your own. If enough of the original remains, copy it. If the handle is missing completely, either copy one that fits your hand comfortably or create your own design. You will need a lathe to turn the end that fits into a socket, fit tang-type handles into their ferrules, or turn the upper ends of hooped handles.

If you wish, you can make an entire handle in a lathe. However, round handles pose a problem, especially those on narrow chisels. They allow the tool to roll, and it could fall off the bench. Some chisel handles are oval, and others octagonal; use handles with these shapes on narrow chisels, as they will not roll easily.

Most chisel handles are made of beech or ash, as these species are shock-resistant. Some "best"-quality chisels have handles made of boxwood or applewood. Both are very dense and take a high polish. You will probably not find any of these species at the local lumberyard, but they can be ordered through specialty lumber dealers who advertise in most woodworking magazines.

The firewood I buy each season always has ash and beech mixed in it. When stacking the wood, I sort out pieces that have straight grain. I then split them into eighths, remove their bark, and set them aside to dry. This way, I always have clear, straight-grained pieces for such projects as chisel handles. When I need to make a handle, I further split these eighths into lathe billets.

When turning a tapered end on a socket-type handle, test the taper's fit. It has to make contact with the wall of the socket along its entire length. If it is too wide at either the bottom or the top, its ability to self-lock will be considerably reduced, or not work at all.

It is more difficult to replace the handle on a tang-type chisel than one on a socket-type. Locate the center of the blade-end of the handle (the mark made by the tail stock center will do just fine). Drill a hole for the tang, being careful to drill along the centerline. If you wander from the centerline the handle will be cocked rather than in line with the blade. When drilling this hole, use a bit with the same diameter as the mid-point on the tapered tang. (See Illus. 337.) This way, the fit of the handle on the tang will be tight, but not so tight as to split the wood.

Before driving the handle onto the tang, first slip the ferrule over the round end. Grip the blade

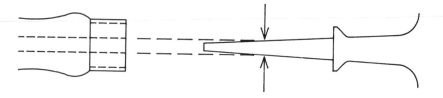

*Illus. 337. When mounting a new handle on a chisel with a tang, drill a pilot hole that has a diameter that's the same size as the thickness of the tang, halfway along the tang's length. The tang and handle will fit tightly, but the handle will not split.*

in a vise so that its bolster is resting on the top of the jaws. (See Illus. 338.) This way, the blade cannot slip. Place the handle on the tang and, with moderate blows, drive it with a mallet until it butts against the bolster.

If you are replacing a lot of handles, you might want to locate a supplier who will sell ferrules. However, for one chisel (or only several) it might be easier to make your own from brass pipe purchased from a plumbing supplier.

Cut the ferrules with a tube cutter. This will result in a shallow lip, or burr, on the inside of each cut. Brass is soft, and this burr can be easily scraped away with a knife burnisher, the burr removed found on most tube cutters, a round file, or even a dowel with a piece of sandpaper wrapped around it.

*Illus. 338. When mounting a new handle on a chisel with a tang, grip the blade in a vise. The bolster will prevent the blade from slipping while the handle is being driven onto it.*

You will eventually drive a mortise or firmer chisel with a mallet. When repairing one of these chisels, remember that you can increase the handle's life expectancy by placing a leather washer between the handle and the bolster. (See Illus. 339.) This will act as a shock absorber. Cut the washer and punch a hole in its center. Slide the washer over the tang, and then mount the new handle.

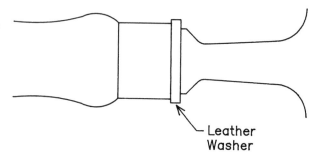

Leather Washer

*Illus. 339. If you drive a firmer chisel with a mallet, it will have to withstand powerful blows. A leather washer between the handle and the bolster will cushion the blows and protect the handle.*

# Sharpening a Chisel or Gouge

Unless a chisel is stored in a manner that will protect it, the exposed cutting edge will inevitably be damaged. Most dealers usually store the tools loose with other tools. As a result, almost every second-hand chisel needs to be ground in order to get beyond the nicks and dings in its cutting edge.

The cutting edges of firmers and paring chisels are usually ground at a right angle to the sides of the blade. This is easier to do on a narrow chisel than on a wide one. To ensure that the cutting edge is both straight and square with the sides, do the following: Spray the back of the blade with layout fluid. Lay the handle of a small try square against the chisel and trace a line in the dried fluid. (See Illus. 340.) Any chips or flaws in the chisel's edge should be on the far side of the line so that they are completely eliminated by the grinding. Grind carefully to the line, quenching often to avoid burning the edge.

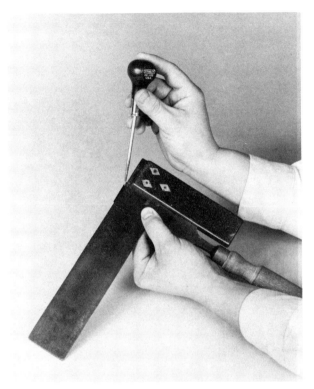

*Illus. 340. A chisel's cutting edge should be square to the sides. If it isn't, use a square to trace a line in layout fluid and grind to this line.*

Remember that the tool is usually hardest near the cutting edge, and becomes progressively less hard (and consequently, tougher) near the handle. If you have to grind away a lot of metal, you may find that the chisel will not hold an edge. You can heat-treat the tool using the information in Chapter 6.

When you are through grinding, take the chisel to the lapping table and flatten the back. Remove any rust pitting as well. Polish the back on a fine whetstone following the information given in Chapter 8. Hone the chisel on whetstones as you would a plane blade. (See Illus. 341.) You can prevent the edge from rolling and maintain the sharpening angle longer if you remember to ride the bezel on both its heel and edge.

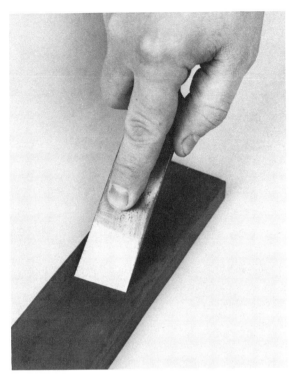

*Illus. 341. Hone a chisel as you would a plane cutter.*

The bezel on a mortise chisel is ground to one of three different shapes: concave, straight, or convex. (See Illus. 342.) The bezel shape is a matter of preference. For mortising in soft wood, use chisels that have a straight or hollow bezel because you do not have to do heavy pulling to break loose the waste. It is possible to break the tip off a mortise chisel, so when working hard wood use a convex bezel. This shape has a thicker and stronger cross section.

Sharpen a mortise chisel as you would a firmer or paring chisel. First, flatten and polish the back. When honing a mortise chisel with a convex bezel, only polish the last ⅛ inch. The rest of the convex shape only adds strength to the edge and does not do any cutting.

Gouges take more time to sharpen than do chisels, but the technique is not very different. If your gouge is sharpened out-cannel, flatten and polish the inside of the curve. Obviously, you cannot do this on the lapping table. If the surface is free of blemishes, it can be polished with a slipstone that has a diameter that's smaller than the gouge's. I use either my round- or teardrop-shaped ceramic file. You can work the slipstone in

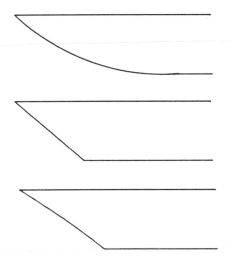

*Illus. 342. Depending on the hardness of the wood you work, grind a mortise chisel bezel to one of the three shapes shown here. The convex bezel at top is the strongest one and can be used on hard wood. The middle shape is straight, at about 35 degrees. The bottom bezel is the same angle, but is slightly concave. This is satisfactory only for soft woods.*

any of several ways: figure eight, side to side, or back and forth.

To remove pitting, cut a short length of dowel that's slightly smaller in diameter than the gouge. Wrap a piece of emery cloth around the dowel and lay it in the curve of the chisel blade. (See Illus. 343.) You will have to work forward and back. This is a slow process and will take patience. When the pitting is gone, polish the gouge with a slipstone as just explained.

Next, hone the bezel. Some woodworkers prefer to hold the gouge and work the stone over the bezel. I prefer to place the stone on the bench and move the tool. This gives me better control.

When honing, rotate the bezel on an out-cannel gouge slowly. Remember to keep both the heel and edge in contact with the stone. When you think you are done, examine the bezel in bright light. You should be able to see the sheen that has developed during the honing. The sheen should be uniform along the entire bezel and should continue right out to the edge.

A light buffing will remove any burr that has developed. Do not overbuff. This will round the inside surface where it meets the cutting edge. After all this work, you do not want to create a problem that you will have to treat in the future.

If your gouge is ground in-cannel, you have to flatten the outside of the curve. If there is no rust or pitting, this can be done on a sharpening stone. Otherwise, take the gouge to the lapping table. As you work the outside over the emery cloth, continually rotate the blade. Next, polish the outside surface on the stone, still rotating the tool.

Hold the chisel on the bench or against your stomach and hone the bezel with a slipstone. Be very careful not to accidentally brush your hand against the edge. Work the slipstone over the entire length of the bezel. You can use any stroke you like: circle, figure eight, side-to-side, or back-and-forth. Make sure that the slipstone rides on the heel as well as the edge. Examine the bezel closely to be sure that you have polished right out to the edge. Finish by buffing lightly.

If you cannot find an in-cannel gouge of the size you need, buy an out-cannel gouge of that size and adapt it. Begin by grinding away all traces of the old bezel. Turn the chisel over and grind a new bezel on the inside. It helps if you have a grinding wheel that has a rounded surface. If not, do as much as you can with the corner of the wheel. Finish shaping the bezel with a round file, such as a chain-saw file. Hone as already described.

# Using a Chisel or Gouge

Chisels are very simple tools that are easy to describe. However, they have an infinite variety of uses, and it is almost impossible to explain them all. They are not commonly used to cut with the grain, as the cutting edge of a chisel has no features such as a mouth to give it control. Thus, the tool tends to follow the wood's grain and can quickly do a great deal of damage. As a result, cutting with the grain is more often done with planes, spokeshaves and drawknives, which are easier to control.

When you use a chisel across end grain, you are *paring*. This is done by holding the tool with the bezel facing upwards. This low angle will give you more control and a cleaner surface. (See Illus. 344.) You will get the best results if you also try to use a slicing action. Do this by holding the edge

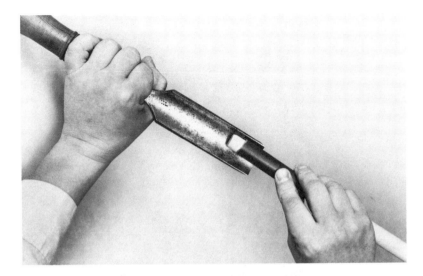

*Illus. 343. Remove pitting from the inside of a gouge by wrapping a piece of emery cloth around a dowel and rubbing it against the blade.*

*Illus. 344. Paring a stop dado with a paring chisel. When paring end grain or across grain, you get a cleaner cut if the chisel is held askew. Of course, in a dado that is not possible.*

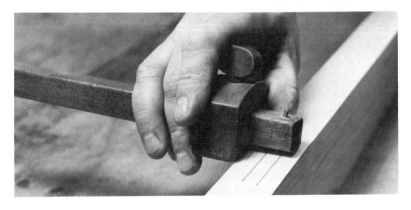

*Illus. 345. Laying out a mortise with a mortise gauge. The lines have been darkened so that they would show better in the photograph.*

at an angle to the direction of the cut. This is called a skewed cut, and its action is similar to that of a plane with a skewed iron.

Before using a *mortise chisel*, first lay out the mortise. This is usually done with a mortise gauge. (See Illus. 345.) This tool is similar to the marking gouge described in Chapter 12. However, the marking gauge has one scribe; the mortise gauge has two. The scribes are set apart the same distance as the width of the chisel. The fence is then set to the distance between the edge of the mortise and the edge of the wood. When the marking gauge is drawn along the wood, it scribes two parallel lines. Use a try square to mark the ends of the mortise. The result is a rectangular box that is an outline of the mortise.

Set the chisel's cutting edge within the scribed box; make sure that it does not overlap the lines. Start near one end of the mortise, but not right at the end. Turn the chisel so that the bezel is facing towards the middle of the mortise. Hold it steady and keep it vertical. Strike the handle with a mallet. (See Illus. 346.) This will drive the edge into the wood. When you withdraw the chisel, you will see a wedge-shaped cut. Move the edge about ¼ inch towards the center of the mortise and repeat the process. Do this several more times, moving in increments of about ¼-inch. (See Illus. 347.) Stop as you approach the other end of the mortise. Leave some waste at both ends until you finish excavating.

As you make each cut with the chisel, some of the short segments being created will break loose. Those that remain in place can be easily levered out. The result is a shallow inlet with the same outline as the mortise. The inlet should have crisp, square corners.

Repeat this process, cutting steadily deeper. Only now, do not pull the chisel free from the wood. After each cut, pull back on the tool so that the edge levers sections of wood loose from the mortise. These sections look like square plugs. (See Illus. 348.) Standing on them is uncomfortable, so when cutting mortises, sweep the floor often.

Do not cut right up to the ends of the mortise. This way, when you pull back on the chisel you will crush waste wood rather than distort the ends. If the ends were to become distorted, the mortise might not be completely covered by the tenon's shoulders. This is not good-quality work.

After cutting several layers of waste, you will reach the desired depth. Now you can square each end by making a single plunge cut. (See Illus. 349.) In soft wood, you can drive the chisel by hand. In very hard wood, you might have to use a mallet.

*Gouges,* which are used for excavating, are often used with the grain. (See Illus. 350.) When coping a concave curve with an in-cannel gouge (or a convex curve with an out-cannel gouge), you make plunge cuts. Be careful to hold the tool at a right angle to the wood. This will shear the wood, and leave a clean cut with vertical edges the same radius as the gouge.

*Illus. 346. A mortise chisel is usually driven with a mallet.*

*Illus. 347. When using a mortise chisel, first make a series of shallow cuts. Note that the cuts do not reach the ends of the mortise.*

*Illus. 348. When you make deeper cuts, you can pop loose large chips.*

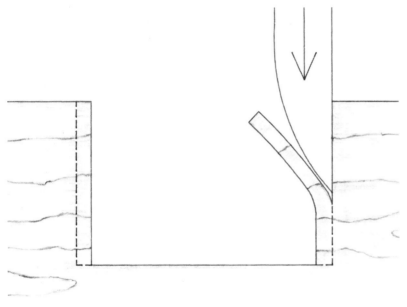

*Illus. 349. After completely excavating a mortise, finish by squaring the two ends with plunge cuts.*

*Illus. 350. Excavating with a gouge.*

# 18
# Saws

When woodworking, I rely almost exclusively on handsaws for a couple of reasons. Perhaps the most important is space. My shop, like many others, is small, and the room required for a table saw is too precious to sacrifice for a tool that I do not often need. My handsaws hang on a wall and take up very little space. Also, five or six good-quality handsaws cost far less than a good table saw.

This does not mean that you (or most other woodworkers) do not need a table saw. Such a decision depends on your individual circumstances. In my case, my commissions are usually limited to single pieces; even large sets of chairs are made only two at a time. When ripping just one or two boards, I find that it is just as quick to snap a chalk line and rip by hand as it is to clean off the table saw (in a small shop clutter accumulates on any flat surface), set the rip fence, and make the cut. A sharp handsaw will cut speedily, and with a little practice you can achieve surprising accuracy.

Think critically about your own woodworking, and you may discover that you too can do all that needs to be done with handsaws. They are very useful tools, and in a small shop where fine work is done they will prove to be essential.

Even though the table saw has been used since the early 19th century, handsaws have always been indispensable tools in most branches of woodworking. As a result, they are common items on the second-hand market, and you can expect to find many of them. You will also see a greater variety of second-hand saws than is sold today.

Before buying any saw, you should understand how these tools work. This information is very basic, but because today most woodworkers rely on either a table saw, Skil Saw, or radial arm saw, the basics are no longer widely understood.

Every handsaw has a metal blade with a serrated edge. These triangular serrations are called teeth. The spaces between the teeth are called gullets. Some saws have larger and not as many teeth. Others have smaller teeth, but there are more of them. A saw is identified by the number of teeth it has. You can ascertain this number by counting the number of pointed tops in an inch of the blade's length. However, saws often have a number stamped into the blade just under the handle, and it is easier to first look there.

There are saws with as few as five points per inch, and some with as many as 20. The fewer the number of points, the faster, but rougher, the cut. The more the points, the slower, but finer, the cut.

*Illus. 351. The rear teeth are from a ripsaw. The front teeth are from a crosscut saw.*

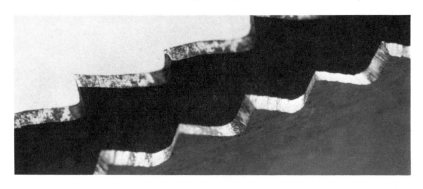

A saw with a low number of points (large teeth) will work well on soft wood, but is more difficult to use on hard wood. Therefore, saws used by carpenters will usually have more points than those used by joiners and cabinetmakers.

Wood is made up of tiny fibres that run roughly parallel to each other. When it is examined under a microscope, it resembles a cluster of drinking straws. The cutting action across the straws (grain) is very different than that along the length of the straws. Just as planes that work with the grain (bench planes) are different from those that work across the grain (skewed rabbets, fillisters and dado planes), saws that cut across the grain are different from those that cut along it. The two different cutting actions of saws are called crosscutting (across the grain) and ripping (with the grain). The shape of the teeth determine whether the tool is a crosscut or a rip saw. (See Illus. 351.)

A saw tooth is triangular and has two angled sides. The front and back edges of a tooth on a crosscut saw are filed at angles to the length of the blade, usually around 60 degrees. This results in two bevelled sides that meet to form a knife point. If you look down onto the teeth, you can see these points. When you look at the teeth from the side, you can see the bevelled edges. (See Illus. 352.) When you are cutting cross grain with a crosscut saw, the row of knives severs the wood fibres, producing fine, powdery sawdust.

A ripsaw tooth is filed at a right angle to the length of the blade. Its front and back sides meet at the top of the tooth to produce a chisel edge. When you look down onto a ripsaw, you essentially see a row of tiny chisels. As the saw cuts along the grain, the teeth actually pare small shavings. If you pick up some of the sawdust created by a sharp ripsaw and look at it in a bright light, you will see that these tiny parings

are very different from the powder produced by a crosscut saw.

When a saw is used, its teeth cut a fine slot in the wood, called a kerf. If the kerf is only as wide as the blade is thick, considerable friction is caused by the blade rubbing against the wood. This not only makes the saw more difficult to use, the wood can actually grab the blade. This is called binding.

Obviously, the kerf has to be wider than the blade's thickness. This is done by giving each tooth a slight outward bend, called the set. Each alternating tooth is bent outward in the opposite direction, (right, left, right, left, etc.) to produce two distinct rows of teeth, one on the left side of the blade and the other on the right. These rows are apparent if you look down onto the teeth.

The teeth on both crosscut saws and ripsaws are set. The greater the set, the rougher the cut, so finer saws (fewer points per inch) generally have less set.

Most handsaws you will find on the second-hand market will belong to one of three categories: handsaws, backsaws, and bow saws. These categories are discussed below.

# Types of Saws

## Handsaws

Handsaws are the type of tool that the word "saw" usually brings to mind. They have a long, wide blade that is stiff and does not buckle when used. These saws are generally used for cutting lumber to dimension. Crosscut handsaws are used to shorten boards, while ripsaws make boards narrower.

*Illus. 352. At left are ripsaw teeth, and at right crosscut saw teeth. 1 shows what you will see when you look down a saw blade. 2 is the view from the side and slightly above the blade. 3 is a section of saw blade that shows the set. The dotted lines represent the width of the kerf. Note that rip teeth have a chisel edge. A crosscut tooth ends in a sharp point.*

Handsaw blades range from those that are as short as 12 inches to those that are 30 inches long. They are always designated by an even number of inches (even though the length may not measure to precisely that number). Smaller handsaws, those between 12 and 22 inches long (12, 14, 16, 18, 20, and 22 inches), are called panel saws.

The next classification of handsaw sizes may cause some confusion. Manufacturers designated their larger handsaws as "hand" saws, so the general group of tools discussed in this section is called handsaws, and a subgroup within this general category is also referred to as handsaws. These handsaws are designated as those that have blades that are 24, 26, 28, or 30 inches long.

Handsaws (the general heading) were made as ripsaws and crosscut saws. However, ripsaws were only made in 26, 28, and 30-inch lengths. As a result, some woodworkers call saws between 26 and 30 inches "ripsaws." This is incorrect, as ripping and crosscutting are a function of the teeth, not the length of the blade. Both ripsaws and crosscut saws were made from 26 to 30 inches long. Saws shorter than 26 inches were generally filed to crosscut.

Ripsaws usually have 5, 5½, 6, 6½, 7, or 8 points per inch. Crosscut saws have from 5 to 12 points per inch. On the second-hand market, you will easily find the most common saws: for example, a 26-inch, 8-point crosscut saw or a 28-inch, 5½-point ripsaw. Fine panel saws (crosscut saws with more than ten points and ripsaws with more than 7 points) are not as common, and you can spend a lot of time looking for them. Instead, follow the advice given on pages 226 and 227 and also find a saw-sharpening service that can cut new teeth.

The lower front corner of a saw blade is called the toe. The lower rear corner is the heel. On some saws, the heel is a convex curve. This is supposed to help reduce damage if the saw jumps out of its kerf and the heel catches on the wood. On other handsaws, the heel is square; sometimes the corner is even clipped at an angle.

The edge opposite the teeth is called the back. Many saw blades have a straight back. Straight-back saws sometimes have a pronounced drop in their forward ends. Sometimes, this drop also includes a small projection about the size of a single tooth. This is called the "nib." (See Illus. 353.) Some have suggested that this nib was used to test the steel's temper. Others believe that it was used to nick the wood to start a saw cut. According to a booklet originally published by Henry Disston & Sons in 1915, the nib is purely decorative.

Some longer saws were made with a shallow, concave-curved back. They are called skewback saws (also hollow-back or sweep-back). (See Illus. 354.) The skewback saw was introduced in 1874 by Henry Disston, the founder of Henry Disston & Sons (at one time the largest saw manufacturer in the world). Supposedly, saws were made with skewed backs to reduce the weight of longer saws and improve their balance. It is hard to determine if this is really the reason. Panel saws as short as 18 inches were made with a skewed back, suggesting that the shape was chosen for aesthetic reasons. On the other hand, skewback saws were generally the most expensive.

Some skewback saws have concave back edges, as well as teeth that form convex curves. Saws are said to be breasted or bellied.

Henry Disston introduced another feature that can often be found on the higher-quality second-hand saws made by many companies: a taper-

*Illus. 353. A nib on a panel saw. The nib is only decorative.*

Illus. 354. Both these saws have skewed backs. The lower saw is also breasted, or bellied—that is, its teeth form a convex curve.

Illus. 355. At far left is a 14-inch panel saw with an open handle. It has been specially filed as a 14-point ripsaw and is used for ripping very thin hardwoods. It was made by Wm. McNiegle, Philadelphia, Pennsylvania. At second from left is an 18-inch, 9 point crosscut panel saw by Groves & Sons, Sheffield, England. In the middle is a 24-inch, 8-point crosscut saw with a label from American Hardware and Supply Co., Pittsburgh, Pennsylvania. At second from right is a 26-inch, 10-point crosscut saw by H. Bishop Co., Cincinnati, Ohio. At far right is a 28-inch, 5-point ripsaw by E. C. Atkins & Co., Indianapolis, Indiana.

ground blade. This is a saw blade that is thicker at the teeth and thinner along the back. When a blade is taper-ground, the teeth require less set to prevent binding. This means that the saw cuts a narrower kerf and requires less effort to use.

These taper-ground and skewback saws were of the "best" quality, and very often their blades were given a mirror polish. If these saws are well-maintained, the blades will remain shiny and free of rust after seven or eight decades. Saws of this quality demand a premium price on the second-hand market, but are fine tools and do excellent work. They are worth looking for and owning.

Saw makers usually stamped identifying information onto the left-hand surface of the blade about equidistant from both ends. Though the stamp could consist of as little as the maker's name, some stamps were very elaborate and contained a lot of information. This, for example, is what was included on Disston skewback saws ca. 1910: the Disston logo, a keystone (the factory was known as the Keystone Works), was centered on the blade. Inside the logo was written: "Established 1840; Henry Disston & Sons; Pat. June 23, 1874; Philada. U.S.A." Next to the logo was this text in italic script: " 'For Beauty Finish and Utility this Saw cannot be Excelled' Henry Disston."

The stamp on a saw blade often revealed the type of steel that had been used to make it. The steel used included cast steel, crucible steel (the same thing as cast), refined crucible steel, London spring steel, and extrarefined London spring steel.

The handle of a handsaw is attached to the blade with a special fastener called a saw nut. These nuts are usually made of brass, and were sometimes nickel-plated. Saw nuts have a wide, flat head (about 9/16 inch in diameter) and a threaded shank. The head is slotted so that it can be tightened with a screwdriver. Some nuts have a broken slot (two small slots, one on each side of the nut). (See Illus. 356.) They have to be loosened with a special screwdriver that has a gap in its blade.

*Illus. 356. These saw nuts have broken slots and can only be loosened or tightened with a special screwdriver that has a gap filed into its bit.*

The saw nut is inserted into one side of a hole in the handle. This hole is countersunk so that the surface of the head just barely projects above the surface of the handle. The threaded shank passes through a hole in the blade; the nut is screwed on from the other side. The nut is the same diameter as the head, and it too is seated in a countersunk hole.

Saw handles were secured with anywhere from two to five nuts. The more nuts used, the more securely a handle was held, and generally the better the quality of a large saw. There is just not enough room on a small saw for five nuts, and in

this case the number chosen seems to have been influenced by the tool's size.

Usually, the head of one saw nut is much wider than the others (an inch or more). Embossed into the surface of this head is the trademark, name, and/or logo of the company that made the saw. Some of these saw nuts are as detailed as a coin; this can enhance the tool's value to a collector. (See Illus. 357.)

Handsaw handles were originally made of wood, often of beech. Later, some handles were made of oak, and during the latter half of the 19th century companies such as Disston began to make the handles out of finer woods such as apple.

During the early 20th century, Disston developed a handle made of an artificial material called Disstonite®. Disstonite® was considered stronger and more durable than wood. The company claimed it could withstand a fall from several stories.

Some newer saws have handles made of plywood, and you will even find some that have plastic handles. I will admit to a strong prejudice in favor of solid wood and will not buy these other saws.

Wooden saw handles were usually varnished, and it is not uncommon to still find them with their original finish. Some saw handles, especially those made of oak and decorated with a vine motif, were also stained a dark color.

# Backsaws

Backsaws are generally used for cutting joints (as opposed to cutting lumber to dimensions). Some of the joints cut with backsaws include dovetails, tenons, and mitres. This work is quite fine, so a backsaw blade is thin and lacks the stiffness necessary to prevent the blade from buckling. For this reason, the blade is reinforced by a metal spine on its back.

Today, many woodworkers cut their joints with a table saw or a router and, as a result, backsaws are not as common as they used to be. There are not as many of them on the second-hand market as there are handsaws. However, they are more common than you might think if your experience with tools is limited to new tool catalogues.

Because they are used for fine work, backsaws have a greater number of teeth; these teeth have less set. Whereas a very fine handsaw would have

*Illus. 357. Some saw nuts are very detailed and can enhance the value of a saw. The saw on the right was made by Ibbotson Peace & Co. Eagle Works. The other, made by Tillotson of Sheffield, England, sports four nuts of graduated sizes. The smaller ones bear England's coat of arms. The largest has a heart and crown separated by an X, designating top-quality London spring steel.*

12 points per inch, a very fine backsaw might have as many as 20. The teeth on a backsaw are filed to crosscut, even though in use the saw sometimes has to cut along the grain.

Backsaws are smaller than handsaws, ranging in length from 6–18 inches. They come in different sizes. (See Illus. 358.) Each size is identified by the purpose for which it was once most commonly used. The smallest backsaw is called a dovetail saw. These saws are usually from 6 to 10 inches long and have from 14 to 20 points per inch. They are used for cutting the dovetails used to join drawers and small boxes.

Many dovetail saws have closed handles. However, because these tools are so small many woodworkers cannot fit their hands into a closed handle. As a result, some dovetail saws are fitted with a pistol grip. Still others have a turned handle that is mounted in line with the tool's spine.

The next size of backsaw is the carcass saw. The large box that makes up a piece of furniture such as a chest of drawers or a desk is called the carcass. These, too, are often held together with dovetails, although they are larger than those cut for drawer construction. A carcass saw was used for this purpose. Carcass saws range from 10 to 14 inches long and usually have 12 points per inch.

A sash saw was presumedly used to cut the tenons and bridle joints used to make sash. The saw is 14 to 16 inches long and has 11 points per inch.

Frame construction (for example, tables and doors) is joined with mortises and tenons. This

*Illus. 358. Shown from top to bottom are the following backsaws: a 16-inch, 10-point tenon saw by W. Butcher, Sheffield, England; a 10-inch, 14-point carcass saw by Henry Disston & Sons, Philadelphia, Pennsylvania; a 10-inch, 14-point carcass saw with open handle by Spear and Jackson; and an 8-inch, 16-point dovetail saw by Hobson of London, England.*

work requires a tenon saw. The blade of this saw is 16 to 20 inches long and usually has 10 points per inch. Because tenons are usually much deeper than dovetails, a tenon-saw blade is up to four inches wide.

The longest backsaws are the mitre saws. They were made for use in a mitre box, and range from 18 to 30 inches long; their blades are 4 to 6 inches deep. The heel is often an S-shaped curve; this ensures that the mitre saw clears the mitre box. (See Illus. 359.) This extends the handle back beyond the end of the teeth, a feature that distinguishes a small mitre saw from a large tenon saw. Although these two types of saws look somewhat different, they can be used interchangeably.

As on their handsaws, saw makers also often stamped information on their backsaw blades. Since these blades are smaller, these stamps are usually less elaborate. A backsaw's spine usually has the company's name and city stamped into it. Some other information (such as the type of steel) may also be included. Owners may have also often stamped their mark on the spine and on the handle.

Most spines are made of iron or steel. On some of the better saws, the spine is made of brass. The brass is attractive, although it does not make the tool work any better. A brass spine does, however, increase the saw's value to a collector, and as a result the saw will usually be more expensive than an identical backsaw with an iron spine.

Backsaw handles are usually made of beech, and are attached to the blade by two or three saw nuts. The back is always straight and not intentionally skewed. The teeth are not breasted, and the blades are not taper-ground.

# Bow Saws

A third type of saw that commonly appears on the second-hand tool market is known alternately as a frame, turning, or bow saw. The most common type of bow saw (known to all woodworkers) is the coping saw. Many larger bow saws were used in the past, but they have been replaced by the bandsaw. These saws are still used in Europe much more commonly than in America. (See Illus. 360.)

I still use a bow saw rather than a bandsaw for the same reason I use handsaws rather than a table saw. A bandsaw would take up too much precious space. Also, a 25-inch bow saw can be used just as quickly, a fact that may surprise woodworkers who have never used one of these tools.

Bow saws are characterized by a frame (made of either wood or metal) that stretches a narrow saw blade and puts it under tension. These saws can

**No. 4.   Mitre-Box Saw.**

**Cast Steel Blade, Mitre=Box Saws.   Blued Steel Back. Apple Handle, Polished Edge.**

Blade measures 4 in. under Back.
Length of Blade, as shown by cut, is about 2 in. longer than toothed edge.

| NO. | Length of Blade, inches | 18 | 20 | 22 | 24 | 26 | 28 |
|---|---|---|---|---|---|---|---|
| 4 Mitre-box Saws | Per dozen   .   .   . | 22.00 | 24.00 | 26.00 | 28.00 | 30.00 | 32.00 |
| The same Saw, 5 in. under Back.   28 in. Blade .   .   . | | | | | | | 36.00 |

One-third dozen in a box.

*Illus. 359. A reproduction of a partial page from a Bigelow & Dowse Co. catalogue that depicts mitre saws.*

*Illus. 360. A bow saw. (Photo courtesy of Don Baldwin.)*

be mounted with a wide rip blade, but the advantage of the frame is that it allows the saw to cut curves, which require a narrow blade. When the saw cuts a curve, it will alternately rip and crosscut. The saw does slow down when it is used to cut with the grain (ripping), but it is still fast enough to do the job. I use a large bow saw with a six-point band-saw blade for cutting Windsor chair seats out of two-inch pine, as well as other curved work. (See Illus. 361.)

*Illus. 361. Using a 25-inch bow saw to cut a curve in a piece of red oak. Note that one end of the board is clamped to the table, while the other is being braced by my body.*

Do not use old wooden bow saws. Leave them for tool collectors. In time, their wooden frames become brittle, and wear loosens the joints. Use a new bow saw. If you do, however, find an old one that appeals to you make a copy of it for use in your own woodworking.

# Selecting and Reconditioning a Saw

Before purchasing a second-hand saw, there are a number of considerations to keep in mind. For example, how does the saw feel? The handle should fit your hand. Backsaws are held with only three fingers. The index finger is locked around the top of the handle. (See Illus. 362.)

*Illus. 362. Many woodworkers find that backsaw handles are too small for their hands. In this case, hook your index finger over the top of the handle. This also helps keep the saw from wandering from side to side.*

Your wrist should be in the same position it would be in if you were about to throw a punch. If your wrist is "broken" (forced into a position that is slightly cocked), you cannot use the saw efficiently. (See Illus. 363.)

*Illus. 363. Ripping a board on a pair of sawhorses.*

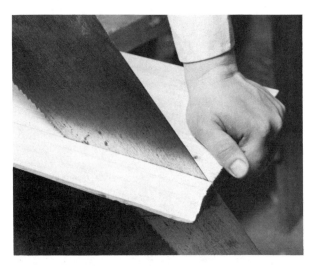

*Illus. 364. When ripping, cut to just the outside of the line so that you leave the line on the work. This way, the piece will not be too narrow. Plane down to the line with a jointer. For short pieces, use a shooting board.*

When a saw is dropped, the weighted end usually hits the floor, so many saws have damaged handles. As a woodworker, you can probably repair a handle or make a new one. However, this is usually more work than it is worth. Still, I have found saws that I liked so much that I have taken the time to repair their handles. When making a new one, use what is left of the old one as a pattern. Factor your time into the equation. I could not justify putting so much effort into a run-of-the-mill saw.

Examine the blade. It should not be pitted, especially near the teeth. Surface rust is no problem, but pitting on the teeth can cause the saw to make a ragged cut.

Sight down the blade, looking along the teeth. This will allow you to see if the blade is taper-ground, which is desirable. You will also be able to see if the blade is kinked or bent. These flaws will make it difficult to cut a straight line. A kink can cause a point on the blade to catch each time it passes through the kerf. Eventually, the saw will bind and kink even more.

Do not buy a saw that has a crack or a break in the blade. Examine the blade carefully, as some cracks are no bigger than a hairline, and can be very difficult to see.

Make sure that the blade has not been greatly reduced in width by repeated sharpenings. This will cause the saw to lose a substantial amount of its weight, so that you will have to work harder. A saw that has been used this way will have very little life left in it. (See Illus. 365.)

Check the saw nuts to make sure that none are missing. They should all be the original ones, and tight. Make sure that the handle is secure and does not rock.

Surface rust can be removed with steel wool (about #1 grade) or 440 grit sandpaper. If you have to remove the handle, remove the saw nuts. If the nuts have a broken slot, you will have to loosen them with a screwdriver that has a gap in its blade. A warding file will do this job.

While cleaning a saw, you may want to polish the saw nuts. Use a brass polish. Use a tooth brush to clean up the large nut with the trademark. Do not buff these nuts, as this process quickly wears away the nut's details.

Grime on a saw handle will usually dissolve in mineral spirits. You can strip a saw handle with paint stripper, but should remember that the original finish increases a tool's value.

# Sharpening a Saw

If you find a saw you want to own and use, do not be concerned about broken, uneven, or badly cut

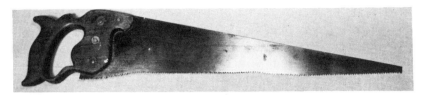

*Illus. 365. This saw was sharpened too many times without being jointed.*

teeth. These problems can be corrected. In fact, you will seldom find a saw that you can take into the shop and use as is. They almost always have to be sharpened.

You can test a saw's sharpness in a couple of ways. Run your fingertips lightly over the points. Next, run them over the sides of the teeth. (See Illus. 366.) In both places, the teeth should catch your skin. If you are careful, you will not be cut.

*Illus. 366. You can tell how sharp a saw is by feeling its teeth. Run your fingertips over the tops and sides of the teeth. If they do not gently snag your skin, they are dull. You can confirm this visually by examining the teeth very closely.*

With a strong light shining over your shoulder, hold up the saw and sight down from the toe towards the heel. Look at the points. If the saw is sharp, you will not see any metal glinting back at you. On a dull crosscut saw, the points will appear as tiny, metal-colored dots. On a ripsaw, they will appear as small metal-colored dashes.

If you are using a saw and hit something hard (such as a nail), sight down the saw from the toe towards the heel; you will be able to see which teeth have been damaged.

As a woodworker, I have a bench grinder and other sharpening equipment, and I sharpen my own edge tools. I will occasionally touch up my saws with a file, especially in the case of a minor accident. However, it requires a lot of time to completely joint, file, and then set each tooth. This is one job I leave to professionals.

A good sharpener will return saws that cut smoothly and evenly, and that do not wander. Unfortunately, a sharpener who does a poor job will send back a saw that is nearly useless. Consequently, the problem is to find a good sharpening service.

No saw sharpener will admit that he cannot do a good job, so the way to be sure of his abilities is to test them. Do this by having him sharpen a common-quality saw. When the saw is returned, sight down the teeth. They should have been jointed, a process that levels the points so that they are all uniform and cut the same.

A good saw sharpener should be able to recut teeth. If you find a saw you like, but it has either too many (or less commonly, too few) teeth, have them recut. Many of my saws have more teeth now than they did when new. Backsaws can also be recut to meet your particular needs.

In my woodworking, I use ripsaws of the following classifications: 26 inches, 7 points; 28 inches, 5 points; and 26 inches, 6 points. My crosscut saws are as follow: 26 inches, 7 points; 24 inches, 9 points; 26 inches, 10 points; and 18 inches, 10 points (panel saw). My backsaws consist of a 16-inch, 10-point tenon saw, a 10-inch, 14-point carcass saw, and an 8-inch, 15-point dovetail saw.

When you are not using your saws, store them in a manner that will protect their teeth. I hang mine on a wall. Many tool boxes have a special case for saws that holds them vertically. Many tool dealers and catalogues sell a special plastic saw guard that slides over the teeth. (See Illus. 367.)

# Using a Saw

If you are going to crosscut a board to a perfect right angle, joint one edge and use a square to lay

out the line. For fine work, use a scribe rather than a pencil, as the scribed line has almost no width. The saw cut should be along the outside edge of the line. (See Illus. 368.)

*Illus. 368. When crosscutting, saw along the outside edge of the line. This way, if you wander from the line, you will not cut the piece so that it's too short. You can make delicate adjustments with a mitre plane, block plane, or shooting board.*

Before ripping, joint the left-hand edge of the board. Starting at one end of the board, measure in from the edge the width of the piece you are cutting, and make a mark. Do the same at the other end. Hook the end of a chalk line at the first mark and extend the line to the second. Draw the line tight and snap it. (See Illus. 369.) Do this by lifting the line slightly and releasing it. This will leave a straight line of colored chalk dust that you can follow when ripping.

Narrow parts can be laid out with a marking gauge (see Chapter 9). This tool is a narrow shaft with a point near one end. Along the shaft is a sliding fence that can be locked in place with a setscrew. Set the scribe by measuring between

*Illus. 369. When ripping wide, long pieces, you can use a chalk line.*

the scribing point and the fence. Hold the fence against the jointed edge and draw the tool along the board. The scribe will create a line that is perfectly parallel to the edge. A panel gauge is a large marking gauge that is used for laying out wide parts. (See Illus. 370.)

A ripsaw leaves a fairly coarse cut. When ripping, stay slightly outside the line, leaving some extra wood that will be removed with a plane.

When you work very thin pieces of wood, there is always a danger they will split; therefore, it is usually best to use a fine saw. A finer saw's teeth are smaller, and they have less set. These characteristics make them less "hungry" and, as a result, less force is required to make them cut. Also, fine saws tend to be smaller in blade length, width and thickness, so they are easier to use on thin work.

*Illus. 370. When you are ripping short, narrow lengths, a marking gauge is the easiest tool to use.*

# 19
# Braces and Bits

The need to make holes in wood is as old as woodworking itself, and drilling tools are some of the oldest artifacts left behind by tool-users. Surprisingly, the bit and brace is a relatively recent tool, being developed only several centuries ago.

Until recently bits and braces were owned by nearly every woodworker, and even by allied building tradesmen such as electricians and plumbers. However, after several centuries of dominance, the bit and the brace have been been replaced by the drill press and the hand-held electric drill in the workshop and on the job site. Since World War II, both machines have become smaller and less expensive and are now affordable by even the smallest professional and non-professional shops.

The drill press and the electric hand drill do many operations very well. The drill press is highly accurate, and the hand drill is fast and handy. These two machines are so inexpensive and easy to use that some woodworkers do not even own a brace. This is unfortunate. Though the drill press and the hand drill have their advantages, so do the bit and brace. Drilling by hand offers more control and sensitivity. A woodworker who can use a drill press and electric drill, but not a bit and brace, is handicapped.

## Braces

Today, whether you buy a brace from a tool catalogue, or at a hardware dealer, there are seldom more than two or three models available. Until several decades ago, tool makers produced a wide range of braces designed for use in the different woodworking trades. Because most woodworkers owned a brace, these tools are still in large supply on the second-hand market. If you do not own one, this is a good place to acquire it.

A brace is essentially a crank. At the top is a dome-shaped disc (usually made of wood) called the head. The head is held by one hand (the left, if you are right-handed). At the other end is a chuck that drives the bit and prevents it from slipping as the brace is turned. The chuck also grips the bit so that it can be withdrawn from the finished hole.

The head and the chuck lie on an axis. In between is the body of the brace, called the frame. The frame is offset to form the shape of the letter C. In the middle of the frame is an elongated handle; this is also usually made of wood. This handle is held by your other hand, and is used to turn the brace on its axis.

The distance between the axis and the offset handle on the frame is called the swing. The swing is the radius of a circle around whose circumference the offset handle travels as the brace is turned. The diameter of that circle (twice the swing) is called the sweep.

Braces are graduated based on their sweep. They usually increase in two-inch increments (which means that the amount of swing increases by one inch.) For example, a brace with a five-inch swing is said to have a 10-inch sweep. In their 1910 catalogues, both Sargent and Biglow and Dowse list a brace with a four-inch sweep (two-inch swing). I have never seen one this small. The largest brace has a 14-inch sweep.

The development of the brace parallels that of the plane. They were first made of wood, often by the craftsman himself. As toolmaking developed into a craft, and then an industry, wooden braces continued to be made. This tradition survived well into the 20th century. As with wooden planes, manufacturers continued to produce and sell wooden braces long after the metal brace had been developed and perfected.

Many of these factory-made wooden braces can be found on the second-hand tool market. Do not use them in your woodworking. Leave them for the tool collectors. These braces were never intended for use with larger-sized bits (above approximately ¾ inch). Large holes were drilled with T-handled augers. Also, over time, the wood these braces are made of becomes brittle, so they often cannot withstand the torque necessary to turn even a smaller bit. Finally, these braces have a brass chuck that is not adjustable. However, you may decide to buy a wooden brace for other reasons. I have a couple that hang in my shop, and I do enjoy looking at them.

During the early 19th century, tool manufacturers began to produce iron braces that were stronger than those made of wood. After mid century, many different designs of iron braces were developed and continued to be sold well into the early 20th century. Many of these iron braces are still for sale on the second-hand market.

The chucks on these earlier types of braces (both wood and iron) posed a problem. Almost all relied on a square, reverse-tapered opening to drive the bit. A thumbscrew or a spring-loaded catch was used to keep the bit attached to the brace when it was being withdrawn from a hole. These chucks were not universal. When a woodworker bought a particular manufacturer's brace, he also had to purchase that firm's bits, as the tapered tang used by other manufacturers did not necessarily fit.

Some of these braces work very well, but owning and using them means that you either have to find bits with tangs that fit the chuck or alter the tangs of the bits you do own. The one exception is the Spofford brace with its distinctive split chuck. This tool was patented in 1859, but was still being made until after 1910. (See Illus. 371.)

The chuck on a Spofford brace is made of two halves that are divided vertically. The halves are connected by a large thumbscrew that is turned to open or close the chuck. This imaginative device was universal in that it would accept any bit with a square tang.

I use a Spofford brace with an eight-inch sweep for making Windsor chairs, which require as many as 45 holes, all drilled by hand. I prefer this tool because when I bore most holes, I lean over the work, and a larger brace bumps against my chest. Also, I have to change bits often, and the split chuck works very quickly and easily.

Spofford braces are common on the second-hand tool market. They, too, were made in a range of sizes, the smallest being a 7-inch sweep (3½-inch swing). They were also made with 8-, 10-, 12-, and 14-inch sweeps. Spofford braces were made with both iron and wood (usually cocabola) heads. The handle on the frame is also wood. It is split (made of two pieces) and is held together with two lead rings.

The modern ratchet brace with its universal, two-jaw chuck was patented in 1864 by an American named William Henry Barber. In 1865, a ratchet mechanism was added. Although this brace competed with a large variety of other types

*Illus. 371. This Spofford brace has an 8-inch sweep. I use it when making Windsor chairs.*

for more than 50 years, it was the eventual winner. As with Bailey's metal planes, this brace was so effective that it has not been greatly improved since it was first introduced.

The Barber chuck consists of a cylindrical cover (called a shell) threaded onto the lower end of the frame. The shell contains a pair of jaws that have a V groove along their inside surfaces.

Their outside edges are serrated. To use the Barber chuck, turn the shell counterclockwise; it rises and the jaws are forced open by a spring. (See Illus. 372.) Place the bit's square tang between the jaws and turn it so that the two opposing corners lay in the V groove. Then turn the shell clockwise; this causes the jaws to close and grip the bit. This chuck is termed universal because it will accept a bit with any size tang.

Barber braces also have a ratchet mechanism that allows the tool to be used in tight spaces, such as between joists, where a complete swing is not possible. The ratchet is also helpful in very heavy work since the brace can be turned only in the position that offers the most leverage.

The ratchet has three positions that are regulated by a knurled ring. When the ring is placed in the middle position, it locks the ratchet so that the brace turns positively in either direction. When the ring is turned (usually to the right), the ratchet is locked so that it turns clockwise, the direction of screw augers, and slips when turned counterclockwise. When the ring is turned the opposite direction (usually left), the ratchet is locked in the counterclockwise direction and slips when turned clockwise.

"Best"-quality braces also have ball bearings under their heads. When used, a brace usually has a lot of pressure on it. This pressure, combined with the constant turning, creates a lot of friction and wear. This wear is reduced by the ball bearings, which also produce a smoother action.

# Selecting a Brace

Braces are very common tools on the second-hand market. There is no consistency to their pricing, but you will certainly be able to purchase an old brace for much less than a new one. Be choosy, and only buy one that is in very good condition.

As with most other tools, "best"-, "common"- and "handyman"-quality braces were produced, and you will see all three on the second-hand market. Only buy a brace that is of the "best" quality. This will have a universal chuck, a ratchet, and a ball-bearing head.

When buying a brace, make sure that all the pieces are present. Disassemble the chuck to be sure that jaws are not worn and that the spring that spreads them is still stiff.

Turn the head. If it wobbles, the tool has experienced excessive wear. As you turn it, also listen for a grinding sound which would indicate that the ball-bearing races have worn. When this happens, you may also feel a vibration through the ball-bearing head.

The wooden head and the handle on the brace's frame should not be cracked either from use, dryness, or water damage. The metal parts (which are usually nickel-plated) should not be rusty. The finish on the wooden parts and the nickel plating is often worn. These problems affect the brace's aesthetic qualities, not its performance.

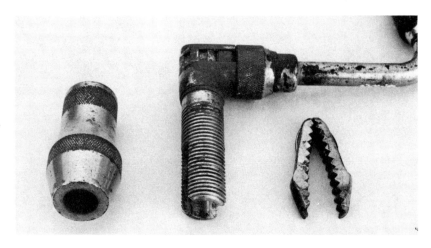

*Illus. 372. A disassembled Barber chuck, showing the shell and jaws. The brace was made by Stanley.*

## Using a Brace

A brace usually has a small oil hole about the diameter of pencil lead under its head. Sometimes there is another hole for the ratchet. These should be given a drop of light oil periodically. If there is no hole for the ratchet, oil the mechanism itself.

# Bits

A number of different types of bits for drilling holes in wood have been developed during this century. However, most of them are intended to be used with either an electric drill or a drill press. These bits have a round shank and no tang. They are not commonly seen on the second-hand market.

Bits intended for use in a brace usually have a tapered, square tang that can be gripped by the chuck. This end is attached to a round or square shank. The differences between the various types of bits occur in the body and the head (the end of the bit that actually bores the hole).

## Augers

The bit you will find in largest numbers is the spiral auger. The first type of spiral auger was patented by Ezra L'Hommedieu in 1809. Additional patterns were developed later in the 19th century. During the early 20th century, augers were used by most woodworkers.

Auger bits have been made for so long and used by such a large number of woodworkers that they are very common on the second-hand market. You can expect to find them easily, and they should be less expensive than new ones.

Although at first glance all augers look alike, they usually belong to three different types. (See Illus. 373.) The first (and the simplest) is the single twist auger invented by L'Hommedieu. These augers are made by twisting a narrow strip of steel around the outside of a cylinder. Later in the 19th century, this same design was improved and sold in America as Ford's patent.

In 1855, Russell Jennings patented an auger that was made by gripping a piece of steel at both ends and twisting it into a spiral. This auger is known as a Jenning's patent or a double twist.

The third auger you will see, and the one most commonly made today, was invented by Charles Irwin in 1884. This bit has a round core running through the body that is a continuation of the shank. The core has a single spiral fin wrapped around it. This type of bit is called Irwin's patent or a solid center auger.

The spiral body of these three types of augers helps carry the chip out of a deep hole. Otherwise, the chip would not eject and would wrap around the bit, causing it to bind. It would be necessary to continually withdraw the tool to clear it.

The body of any spiral auger is the same diameter as the hole it drills. This prevents the bit from wandering in a deep hole. The edge of the spirals that makes contact with the side of the hole is called the web.

Augers that are intended for general woodworking range from about 7½ inches to about 9½ inches long. Augers with smaller diameters are shorter than those with larger diameters. Augers were also made for a large number of special purposes, for example, shipbuilding and railroad car construction. These bits are much longer—from 12 inches to 18 inches. They are occasionally seen on the second-hand market, but are not discussed in this chapter.

The lower end of an auger is known as the head. Ironically, the head on a bit is at exactly the opposite end of the head on a brace when the two tools are assembled for use.

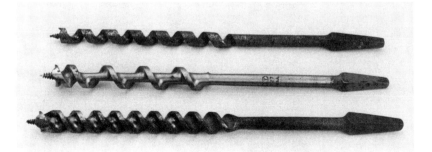

*Illus. 373. The top auger is the single-twist type invented by Ezra L'Hommedieu. The middle is an Irwin patent, or a solid center auger. The bottom auger is a Jennings patent, or a double-twist bit.*

The auger head that is most useful in general woodworking (on the solid center auger, Irwin's #6 head) has several critical parts that enable the bit to cut a hole. The first is a short, stubby screw lead. The lead centers the bit on a mark, and once screwed into the surface of the wood, pulls the rest of the head and the body through behind it. (See Illus. 374.)

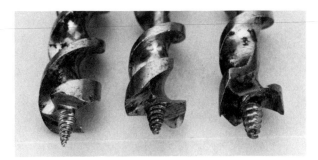

*Illus. 374. A closeup of an auger's head. This auger is an Irwin patent. It has two spurs and two lips.*

Besides pulling the bit into the wood, the threads on the lead also regulate the auger's cutting speed. Just as a nut on a coarse-threaded bolt will travel further in one turn than a nut on a fine-threaded bolt, a lead with coarse threads will travel further in a single turn of the brace than will a lead with a fine thread. Because a coarse-threaded lead will pull the head further each turn, the bit will also cut a heavier chip and take less time to finish the hole. One of these bits with a coarse-threaded lead is said to have a "fast" screw.

You will find augers (or screws) with three different leads. They are called fast, medium, and fine augers. (See Illus. 375.) A fast auger will bore a hole quickly, but (because it takes a thicker chip) requires more effort to turn. It is useful for general woodworking and for carpentry. A fine auger will cut more slowly, but it takes less effort to turn. However, because the chip is thinner, it takes more turns to make each hole. A fine auger cuts best in very hard wood. Medium augers are good for soft wood.

The head on either a Jennings' or an Irwin's patent auger usually has two spurs on its outside

*Illus. 375. Shown from left to right are slow, medium, and fast screws.*

edges, while a single-twist auger has just one. As the lead pulls the head of the bit into contact with the wood's surface (Illus 376), these spurs score the outline of the hole. (See Illus. 377.) This first step is similar to the job done by the scribes on a dado plane. By scoring the hole's circumference, the spurs on an auger prevent tear out of the surrounding wood.

The next part of the head to make contact with the wood is the cutting edge, called the lip. Solid-center and double-twist augers have two lips; a single twist has just one. Lips are inclines (like plane blades) sharpened at their lower ends. As the lead pulls them into contact with the wood, they begin to lift the chip from inside the circle scored by the scribes. (See Illus. 378.)

As the chip is lifted, it is passed out into the

*Illus. 376. When an auger bit makes a hole, first the screw lead will enter the wood.*

*Illus. 377. Next, the spurs score the outline of the hole.*

*Illus. 378. Finally, the lips begin to cut loose a chip.*

first twist, which is called the throat. From the throat, it is passed up into the spiralled body and carried out of the hole. Since Jennings' and Irwin's patent augers have two lips, two chips are cut simultaneously.

While the bit is turning, all four parts of the head are in operation, performing four separate functions all at one time. The lead pulls the bit further into the wood, while the spurs continue to score the hole's outline. At the same time, the lips

cut loose the prescored chips and pass them up into the throat. From the throat they are fed up the spiral, until they reach the top of the hole and are ejected. As long as the bit is turned, it will cut continuous chips that usually break up rather than hold together.

Augers are graduated in ¹⁄₁₆ths of an inch. They range from ³⁄₁₆ to 1½ inches wide and are usually numbered in the same way as plow irons. For example, a #3 auger is ³⁄₁₆ths inch wide, a #4 is ¼, a #5 is ⁵⁄₁₆ths, a #10 is ⁵⁄₈ths, and a #14 is ⁷⁄₈ths, etc. Augers wider than one inch (#16) increase by ⅛ths and are numbered 18, 20, 22 and 24. Some companies made a #17 (1¹⁄₁₆-inch) auger.

The augers available on the second-hand tool market were made by many different manufacturers, but they are all sized in the same way, so a set can be assembled. I use solid-center and double-twist augers, and do not prefer one over the other.

Augers have always been sold either individually or in sets. Manufacturers offered their sets in either a roll or a special wooden box. Occasionally, sets of augers are found still in their original box or roll. (A roll is a long strip of canvas with small individual pockets for the bits.) While collectors are seldom interested in individual bits, a complete set, in its original box, is more desirable. In this case, expect the per-unit price to be substantially higher than the price you would pay for single bits.

The critical cutting heads of augers (as well as all other bits) have to be protected from damage. When I am travelling around the country to a workshop, I carry my bits in a roll. However, when working in my shop I find the roll to be a nuisance. An alternative is a bit block, a block of wood, at least 2 inches thick, with one or more rows of holes in it. The holes should be ½ inch in diameter and should be spaced about 1 inch apart. The tangs of your bits are placed in these holes. (See Illus. 379 and 380.)

Since the bits are stored with their cutting ends pointing upwards, exposed to anything that might fall on them, I keep my block on a middle shelf in a set of shelves. The shelf above the bits protects them from damage. I have a smaller bit block screwed to the wall behind my bench, and in it I keep the bits I use most frequently.

*Illus. 379. A bit block is a practical way to store bits. Make sure that you put the block on a shelf or somewhere else where nothing can fall on the bits.*

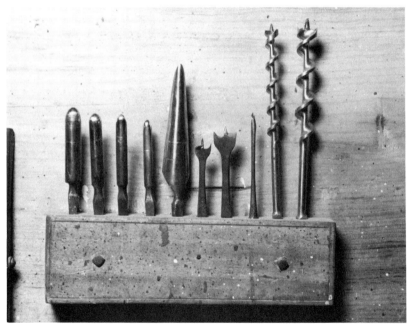

*Illus. 380. A smaller bit block hangs on the wall behind my workbench, where it is within easy reach. It contains the bits that I use most frequently.*

## SELECTING AN AUGER

Before purchasing an auger, check that neither the shank nor the body is bent. This flaw can be detected if you either hold the bit up to your eyes and sight along it or roll it on a flat surface.

Next, check the head. Dealers normally store old augers in a container, loose with other metal objects. You will usually find them on the floor in boxes. This careless storage, plus accidents (such as being dropped on the lead), can do considerable damage.

Check the screw lead. It should be pointed so that it can start and tap its pilot hole. The screw's threads should be well defined and not flattened or crossed. Damage to the threads may cause the lead to strip the pilot hole, with the result that the bit will not pull itself into the wood as it is turned. This would make the auger almost impossible to use.

The spur(s) that scores the diameter of the hole should still have its full form. The spurs on bits that have been sharpened often or that have seen a lot of use have been almost worn away. They cannot score deeply enough to undercut the edge of the chip. The result is a ragged hole, or tear-out, in the surrounding wood. Abuse or careless handling will sometimes break off a spur or roll it so badly that it cannot be repaired.

The lips of an auger you are buying should not be badly nicked or worn by repeated sharpenings. Neither the spurs nor the cutting edges should be pitted by rust.

Augers are so common that you can afford to be very selective when buying them. Do not accept one that has any of the problems mentioned above.

## SHARPENING AN AUGER

Before using an auger, clean off any surface rust. Next, sharpen the bit.

Nicholson File Company makes a special bit file which is sold by some tool dealers and through catalogues. This file is quite thin. It is double-ended, and each end is strongly tapered. (See Illus. 381.) One end has safe sides (see Chapter 8); the other end has safe edges. There are no teeth on the adjacent surface of the bit file to damage other parts of the bit when you use the sides or edges of the bit file. Flat needle files also have safe edges and can be used if you cannot find a bit file. After filing, some woodworkers also hone the lip and spurs with a triangular slipstone.

First, examine the edges of the spurs. Under a bright light, look directly down onto the spurs and pivot the head front to back.

You should not see any light glinting along the edge of either spur. If you do, it is dull at this point and needs to be sharpened. Do this only from the inside. Never touch the outside edges. Doing so would reduce the diameter of the circle they score. The edges of the hole will be ragged and you may experience tear-out.

The lips may have to be sharpened as well. This, too, is done with a bit or a needle file. Place the screw on a piece of soft wood such as pine. Turn it once or twice so that the lead begins to enter the wood. This will prevent the bit from walking as you sharpen it. (See Illus. 382.)

With one of the lips facing you, tilt the bit backwards. File along the lip's upper edge only. If you file the bottom, you may reduce the clearance. With the file, remove any nicks or spots that have

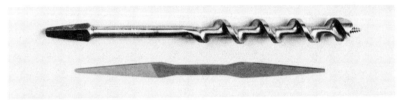

*Illus. 381. A bit file and auger. Note that the left end of the file has safe edges, while the right has safe sides.*

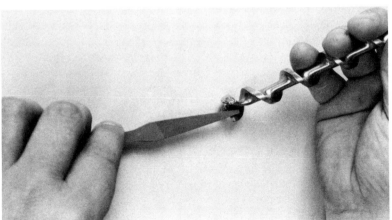

*Illus. 382. Using a bit file to sharpen an auger's lip. The safe edges allow you to file the entire lip without damaging the rest of the head.*

been rolled over. Be sure to balance the amount of filing, so that you remove an equal amount of metal from both lips. Otherwise, one side will cut a heavier chip than the other.

Most woodworkers sharpen their augers much more often than is necessary. Examine a recently acquired auger to determine what care it needs. After that, sharpen only when the bit is dull.

### USING AN AUGER

Begin the hole by setting the lead on a mark. Augers work best at 90 degrees. If you try to drill a hole at an angle, the spurs may not score the complete outline of the hole, and the lip will often

*Illus. 383. When drilling, it is usually advisable to back up the work with a piece of wood.*

*Illus. 384. If you do not drill into a waste block, most bits will tear out the surrounding wood as they exit.*

tear out a chip from the surrounding wood.

When an auger exits from a hole, it also tears wood from the edges. When drilling through holes, clamp a waste block to the other side of the wood. (See Illus. 383 and 384.) The tear out will occur in the waste, leaving clean edges in the wood you are drilling.

# Center Bits

Unlike augers, center bits are no longer being made, and, as a result, most woodworkers are not familiar with them. However, they were sold and used in great numbers from the 18th century until well into the 20th. They were a general-purpose bit used in carpentry, cabinetmaking, and similar trades. (See Illus. 385.)

In America during the second quarter of the 19th century, the center bit replaced the spoon bit in common chairmaking (Windsor, ladder-backs, and Fancy chairs). This change coincided with the transition in chairmaking from craftshop to factory production. Center bits were also used for roughing mortises and for making inlets. For example, gunmakers often used center bits for roughing out the inlets for gun locks. (See Illus. 386.)

Center bits make accurate, flat-bottomed holes. They can also drill clean, crisp holes at angles as low as 45 degrees in both flat and round surfaces. (See Illus. 387.) They were used in hand drilling for many of the same purposes as the modern Forstner bit.

Because they lack a spiralled body, center bits cannot clear a chip from a deep hole. The larger bits are most effective when used for drilling holes with a depth of up to about twice their width. Smaller bits are good at depths up to perhaps three times their width.

Center bits were developed before Barber invented his universal chuck, and were made for use in earlier chucks that had a square, reverse-tapered opening and a spring catch. The shanks of center bits are usually square and are swollen in the middle. The shank decreases in thickness near the bit's head. This creates a narrow neck and leaves more space around the top of the hole for the chip to exit. The upper end of the shank is tapered to create a tang (the end that is placed in the chuck). This end will usually have a notch filed in it for the spring catch, but fits very nicely

*Illus. 385. These center bits range from 3/16 inch to 1½ inches.*

*Illus. 386. The gun maker who fitted a lock in this musket used three different sizes of center bits to rough out the lock inlet. After he used the bits, very little wood needed to be chiselled away.*

*Illus. 387. Even at a steep angle, a center bit will drill a hole with a clean edge.*

into a modern Barber chuck. It is gripped equally well in the split chuck of a Spofford brace.

While an auger has a round cutting head, a center bit's head is made of two flat wings. (See Illus. 388.) Its cutting action is similar to that of a spiral auger, and both heads have three features in common: a lead, spur and lip.

The lead is the first feature to make contact with the wood. It is not threaded and usually has three distinct facets, so that it looks somewhat like the point of a drawn wire nail. One of the facets is flush with the wings. (Some 20th-century center bits have a screw lead, but I have only seen them illustrated in old catalogues.)

On the outside edge of one wing is a spur. After the lead has entered the wood, this spur is the next feature to make contact. As the bit is turned, the spur scores the circumference of the hole the bit is about to bore.

The second wing has a lip (on center bits, it is also called a router), which is raised at an angle

*Illus. 388. A close-up of a center bit's head. Note that the lead is not round; instead, it is made up of three facets.*

and has a sharpened cutting edge. As on a spiral auger, this lip pares the chip out of the hole. This lip only cuts in a clockwise direction.

Like augers, center bits are graduated by sixteenths of an inch. The smallest are ³⁄₁₆ inch, and they range up to 2 inches in diameter. They are not usually numbered, and you have to measure the bit to know what size you are using. This is easy. The wings of a center bit are flat, so you can easily measure its size with a ruler. Center bits are short in comparison to other bits. My ³⁄₁₆ths-inch bit is a little over 3 inches long, and my 2-inch bit is only 5 inches long.

Center bits were used for nearly two centuries in a large number of woodworking trades. As a result, they are very common on the second-hand tool market, and are usually inexpensive.

When new, these bits were sold both individually and in sets. Today, original sets are uncommon. Like all other second-hand tools, over time the bits in sets have become separated. You can acquire center bits individually, buying the sizes you need as you find them. On the other hand, some dealers assemble individual bits and then sell them as sets. Reputable dealers identify these sets as being assembled.

If you are offered an original set, expect them to be more expensive than an assembled set or the same number of bits acquired one at a time. Following are some of the ways you can recognize an original set: One, the bits will all look alike.

Two, since the bits had to fit a single brace, their tapered ends will all be the same size. Three, center bits are occasionally stamped with the maker's mark. If one or more bits in a purported set is marked, they should all be.

Because center bits have a pointed lead rather than a screw, they do not travel a prescribed distance into the wood with each turn of the brace. In other words, manufacturers did not make "fast" center bits. You regulate their speed by increasing or decreasing the amount of pressure put on the brace.

## SELECTING A CENTER BIT

Before purchasing a center bit, examine it for the same defects that can be found on spiral augers. The lead should not be bent or twisted. Check the scribe's height. It has to be long enough to make contact with the wood in advance of the cutting lip. Examine the outside of the spur for any indication that a past owner has filed this edge. If so, the scribe will score a hole that is smaller in diameter than that cut by the lip. The result will be a lot of friction, a hole with a rough edge, and perhaps a chip torn out of the surrounding surface.

Examine the lip. This feature has been periodically sharpened over the years, and it can become worn. A goodly amount of metal should remain on any center bit you buy. The lip should not be bent, damaged, or pitted by rust. Check the width of the lip to be sure that it is as wide as the spur. Otherwise, it will not remove a complete chip. A ridge of wood will be left around the outside of the hole. The bit will soon stall when the inside surface of the curved spur-wing begins to ride on this ridge. (See Illus. 389.)

Center bits sometimes have a twisted shank, caused by careless use. They can be straightened without breaking because only the head is hardened. The shank is left soft, and therefore, tough. (See Chapter 6.) Grip the flat wing in the jaw of a metal vise. You cannot grip the angled lip without damaging it. To avoid the lip, let it project above the vise jaws. This will cause the shank to be angled towards the floor. (See Illus. 390.)

Fit the chuck of a bit brace over the end of the shank and tighten the jaws to obtain a secure grip. The bit is usually twisted clockwise. Making sure the brace's ratchet is locked in the counterclockwise direction, carefully, and slowly, twist

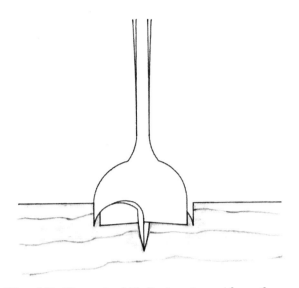

*Illus. 389. If a center bit's lip is not as wide as the spur, a raised ridge will appear around the edge of the hole. This ridge will build up under the scribe and will quickly stop the bit from cutting.*

*Illus. 390. To straighten a center bit with a twisted shank, grip the spur side of the head in a vise, and the other end in a bit brace. Turn the brace very slowly and steadily in the direction opposite the twist. Watch the shank; you will see the twist disappear.*

the shank that way. As you do, watch the corners of the square shank. You will see them straighten. There will be some slight springback when you release the pressure, so just barely overbend the shank.

Center bits can be bent, as well as twisted. A bend can be removed by laying the bit on an anvil (or other flat, metal surface) and tapping it with a hammer.

## SHARPENING A CENTER BIT

As with augers, center bits are usually sharpened more often than is necessary. They should gener-ally be sharpened when they are purchased, and not often after that.

A center bit has just one spur and cutter. Sharpen these as you would an auger's, using a bit file or a needle file. Never touch the outside edge of the spur.

To prevent the bit from walking while the lip is being filed, lay the upturned lip on the edge of your workbench or a block of wood. The shank will be at an angle. As on a spiral auger, the lip is sharpened on the upper edge. When filing the lip, try to avoid contact between the lead and the file (it helps if the file has safe edges).

Any pits or nicks in the lip make the bit less efficient and will greatly increase the amount of work necessary to drill a hole. If the bottom edge is rust-pitted, dress it on the lapping table. The pointed lead will force you to work on the very corner of the lapping surface. (See Illus. 391.) Make sure that you maintain the angle of the lip's underside. If you form a second angle, you reduce the clearance and limit the thickness of chip the lip can cut.

The sharper the cutting edge, the easier the bit is to use. For this reason, you might want to hone the lip. I do this with a triangular ceramic file.

*Illus. 391. Sharpen a center bit router on the square edge of a stone. Note that the upper surface has been lapped, as would be the upper surface of a chisel or plane blade.*

## USING A CENTER BIT

I use center bits in delicate work and in thin stock for the following reasons: One, because the lead is pointed, it causes less disruption of the wood under the hole than does a screw lead. Two, the pointed lead is less likely to split thin stock. Three, I (not the lead) control the speed. Four, because there is only one lip, center bits cut more slowly and precisely than do spiral augers.

Place the point of the lead on the mark that centers the hole. If the hole is to be at an angle, try to start at that angle. Later, after the head has begun to penetrate the wood, you can make some slight correction of the angle by pulling the brace in the desired direction. Do not try to make too much correction, as this can damage the lead. This ability to correct the angle of a hole is an advantage the plain lead has over one that is threaded.

Since this plain, pointed lead does not pull the bit through the wood as a threaded lead does, you have to create downward pressure on the brace, usually with your chest or chin. The amount of pressure applied is a function of the thickness of the wood and its hardness. If the part is small and fragile, bore more slowly and with less pressure.

These bits do not have a spiral body, so they do not eject their chips from a deep hole. When boring a hole that is deeper than the head, retract the bit periodically to clear the chip. Otherwise, the bit could bind and the shank might bend. Also, the friction could produce enough heat to draw the temper.

You might find an old chairmaker's technique handy in your own woodworking. For strength, chair joints have to be as deep as possible. When chairmakers began to use center bits in the 19th century, they usually drilled the hole until the point of the lead just barely pierced the other side. Such a tiny hole was not noticeable and was often filled by the finish. If you ever have to repair a mid-19th-century common chair, such as a Windsor, check to see if a center bit was used. (You can tell by the hole's distinctive bottom.) Hold the part up to the light and you can often see the pin hole that you would have otherwise never noticed.

# Gimlets

Not all the holes made in wood have to have crisp edges and clean sides. Consider, for example, pilot holes for nails and screws. They relieve the pressure caused by the fastener, prevent it from splitting the wood, and keep the fastener from binding, which could cause it to bend or break. For such work, speed is more important than appearance. Gimlet bits were made for just this purpose.

Gimlets, like center bits, were used from the 18th century until well into the 20th century. They are very common on the second-hand tool market and are usually inexpensive. Manufacturers were not able to agree on what to call these bits. In catalogues, they appear alternately as German, Swiss, or Norwegian gimlets.

Gimlets are graduated by $\frac{1}{32}$ of an inch, and are numbered sequentially. They range from $\frac{1}{16}$ inch ($\frac{2}{32}$, which is marked #2) to $\frac{1}{2}$ inch. A $\frac{1}{2}$-inch bit will be marked #16 ($\frac{16}{32}$ inch). Gimlets are shorter than augers, but not as short as center bits. My $\frac{1}{2}$-inch bit is 7 inches long, and my $\frac{3}{32}$-inch bit is 5 inches long. (See Illus. 392.)

Like augers, gimlets are spiral bits, and the purpose of their spiral bodies is to carry the chip out of the hole. However, the similarities between augers and gimlets end here. A gimlet has a very different cutting head. Also, its body is tapered, rather than having a constant diameter.

A gimlet is actually a two-spiral bit, and the bit's cutting action is a two-step process. The first

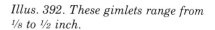

*Illus. 392. These gimlets range from ⅛ to ½ inch.*

spiral is the cutting head, and is located on the very point of the bit. The head twists about two-thirds of the way around the bit. It has a low pitch, so that on the larger gimlets it travels up the body about ¼ inch (about ⅛ inch on the smaller bits). The head is so small you have to look closely to examine it. (See Illus. 393.)

The second spiral is on the body of the bit. Unlike an auger, this spiral has a steep pitch, so that the entire length of the body is just one single twist. The spiral body is also swelled. It increases gradually from just behind the cutting head to its maximum diameter. It then narrows into the shank.

A gimlet's cutting action is unlike that of a spiral auger or a center bit. Those bits score a circle and then lift out the wood inside the circumference. A gimlet's head cuts a narrow, primary hole; the diameter of this hole is then increased by the body. This is a paring action, performed on the sides of the hole by the sharp edges of the metal spiral. The parings fall into the spiralled body and are carried out of the hole.

How well a gimlet works is a function of its diameter and the hardness of the wood. Gimlets of all sizes work best in hard woods. Of course, pilot holes are more often needed when working with hard woods, since these species are more inclined to split when a nail or screw is driven into them. Hard woods are also more likely to cause a screw or nail to bind. A small gimlet used to drill a hole in the soft wood such as pine will act more like a brad awl than a drill bit. It makes its hole by pushing the wood aside, rather than actually cutting it.

Since the top of the hole made by a gimlet has been enlarged by paring (rather than being scored by a scribe), its edges will be slightly ragged. This is not usually a problem. If the hole is a pilot for a wood screw, a countersink will remove the ragged edge. If it is a pilot hole for a nail, the nailhead will cover its edge.

Gimlets cut quickly when used to drill edge grain. In end grain, gimlets cut more cleanly, but more slowly (although still fast relative to center bits and augers).

## SELECTING A GIMLET

When buying a gimlet, sight along its shank to be sure that it is not bent or twisted. Next, look at the tip. The point should be sharp. If the bit has been dropped or abused, the point may be rolled over; this will prevent it from starting cleanly and accurately. Test the point by pressing it into your fingertip. You can quickly feel the difference between one that is sharp and one that is dull.

Next, check its cutting head. The partial spiral should be sharp and well-defined. The head's paring edge should not be damaged. On larger bits, you can see the edge clearly, but on smaller bits it is too tiny to examine closely. Try scraping it across your thumbnail. It should scrape a shaving. If not, it may be rolled over and will not be able to pare wood either. The bit will turn, but not cut.

Finally, look at the edges of the body. They, too, should be crisp, without major dents or nicks.

## SHARPENING A GIMLET

Gimlets are not easy to sharpen. After tuning one, you should use and store it carefully.

Because their features are so fine, small gimlets are more difficult to sharpen than larger ones. A triangular needle file will sharpen the ridge behind the first spiral, where the body begins. Touch up the inside of the cutting head with a round needle file or a small chain-saw file. Start right at the point, under the curl of the first spiral. (See Illus. 394.)

Touch up the inside edge along the length of the body. Keep the file against the inside surface of the throat. If you lift it, you may roll the edge, and the round burr you create (rather than the edge) will contact the wood. The burr will not cut, and the result will be a lot of heat that could draw the bit's temper.

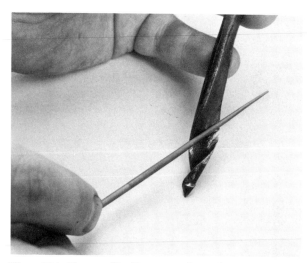

*Illus. 394. Use needle files to touch up the inside edges of a gimlet's body and also on the head.*

When first tuning a gimlet, you may have to hone the outside of the spiralled body to remove rust or nicks from the paring edges. Do this only once. Place a sharpening stone on the bench top and run the gimlet over it. (See Illus. 395.) Find a position that brings the edge into contact with the stone, and move the gimlet back and forth. At the same time, slowly turn the bit so that you hone along the entire length of the spiral. Do as little of this honing as necessary. Never grind or file this edge.

### USING A GIMLET
A gimlet will start on its point. Set the tip of the bit on the mark and turn slowly until the head is buried. Then you can increase speed. (See Illus. 396.)

*Illus. 395. Sharpen the outside edges of a gimlet's body on a sharpening stone.*

Gimlets will drill at very shallow angles, but if you start at an angle you take the chance that the bit will walk across the surface, and perhaps damage the wood. Instead, hold the brace at a right angle, and once again, bury just the head. Now, pull the brace to the desired angle. As you move the brace, you should also be turning it.

*Illus. 396. Using a gimlet.*

## Spoon Bits
Of all the tools that woodworkers use for making holes, perhaps only the awl is older than the spoon bit. The Romans used these bits, and they have also been excavated from Viking sites. Although an ancient tool, spoon bits continued to be made in England until only a couple of decades ago. They are still made by an American tool company.

Ancient spoon bits looked like an open bowl on the end of a round shank, almost exactly like a teaspoon. Those that were made and used in America and England for the last two centuries more closely resemble a narrow test tube sliced in two along its length. In other words, a spoon bit is half a cylinder with an upturned nose, so that the bit is capable of holding water. (See Illus. 397.)

Spoon bits had been in use for centuries before the development of the common chair industry in the 18th century. The craftsman who began to semi-mass produce common chairs such as Windsors, ladder-backs, and Fancy chairs adopted the

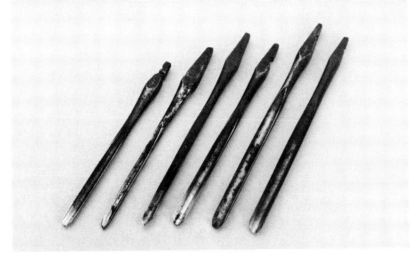

*Illus. 397. Spoon-type bits.*

*Illus. 398. These chair bits range in size from ⅜ to ¾ inch.*

spoon bit because these tools had properties they required in their work. The bits they used became so closely associated with the chairmaking industry that they became known as "chair bits." (See Illus. 398.)

A chair bit is about 5 inches long overall. The body of the bit is about three inches, the rest being a short, thick shank and the tang. Chair bits usually range in diameter from ⅜ to ⅞ inch. When working, a chairmaker might use no more than four different sizes.

These bits are roughly graduated by ⅛ inch, but vary considerably. Since a chairmaker makes a tenon to fit the hole, exact sizes are not necessary. In fact, a bit was designated by its purpose rather than its size—for example, the "leg bit," "stretcher bit," "stump bit," etc.

The earliest chair bits were permanently fixed in the brace, and chairmakers had to own as many braces as bits, perhaps several. Later chair bits were fixed into a wooden mount, called a pad. The pad had a square tang that fit into the brace, so that the chairmaker owned only one brace, but several pads. Later, as chucks were developed, chair bits were made with a square shank. These bits will fit securely in either a Spofford- or Barber-type chuck.

Sets of chair bits are rare (although I was once fortunate enough to purchase one). You generally have to assemble a set. Dealers do not always recognize them, and they are often inexpensive.

Another type of spoon bit was used by other woodworkers. These spoons differed from chair bits in that they were longer and were usually made in smaller diameters. Overall, these spoon bits are usually between 7 and 8 inches long, with the body of the bit about 4 to 5 inches. The bit has a round shank and a square, tapered tang. Since these bits were often used on early braces with a thumb catch, the tang usually has a notch filed in it. However, they fit equally well in a modern brace. These spoons range in size from 1/16 to ½ inch. They are common items and are usually inexpensive, often turning up in the "dollar box."

You will also find another type of bit that is quite similar. These are called alternately, shell, quill, or pod bits. The only difference between shell and spoon bits is the nose. On spoon bits, the nose is upturned, while on shells it is open. In other words, a shell bit will not hold water. It is possible that a shell started out as a spoon bit and lost its nose to repeated sharpenings. On the other hand, shells were also made in great numbers. You will see many of them on the second-hand tool market, and they are inexpensive.

There are many other variations of spoon-type bits, but I am not including them in this section.

Spoon and shell bits work similarly. As they

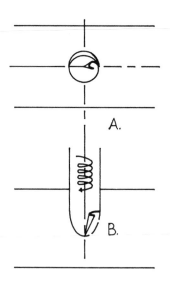

*Illus. 399. A shows a spoon bit paring a shaving from the bottom of a hole. B is a cutaway view of the bit's lip cutting a shaving, which is actually a spiralled plug.*

are turned, the nose's curved edge slices a chip from the side of a round bottomed hole. (See Illus. 399.) The cutting action is continuous, so the chip is held together to form a tight core. When a spoon is in use, the chip rises up the hollow body; it looks like a cork being withdrawn from the mouth of a wine bottle.

Spoon, chair, and shell bits have a number of advantages that are not combined in any other type of bit with which I am familiar. Because these bits do not have a screw lead, they do not cut at a predetermined rate of feed. The bit's speed is instead regulated by the amount of pressure you apply to the brace. You can cut very slowly and gently or (in a situation such as making a Windsor chair, when you have to bore as many as 45 holes), you can apply great pressure and achieve surprising speed.

Remember, the noses of spoon and shell bits cut a chip from the side of the round bottom hole. When a through hole is drilled, the point of the nose emerges first. The exit hole is then gently opened to the bit's full diameter, resulting in less breakout than with other bits.

Spoon bits cut very well at an angle, a feature that is important in hand chairmaking. Furthermore, because they do not have a lead, you can actually change the angle after the hole has

begun. Since many Windsor chairmakers drill holes by eye, this is an important feature.

### SELECTING A SPOON BIT
Do not buy a bit that is bent or badly pitted from rust. Otherwise, spoon bits are very simple tools and usually (even though they have been mistreated) not much damage is done to them. Even if the nose of a spoon bit were snapped off, it could still be resharpened as a shell bit.

### SHARPENING A SPOON BIT
Spoon bits present a unique sharpening experience. A turning gouge is the only tool I can think of that is even remotely similar. The cutting is done entirely by the edge of the bit's nose, and this profile is all that needs to be sharpened. More specifically (if you turn the brace clockwise), the cutting is done by the left half of the nose (looking at the open side of the body). The edges are not sharpened, and only serve to keep the bit from wandering.

Do as little grinding as possible to either a spoon or a shell bit. You should only grind to make the outline of the nose symmetrical. The proper shape for the nose is the pointed end of an ellipse, rather than half a circle. There is no bezel. The outside surface curves smoothly upwards to form the nose. Once you have established this shape, do any future sharpening by honing. I have chair bits that I have used for 17 years without ever grinding.

Next, hone the nose's cutting edge. If a shell bit is held at about 35–40 degrees, its cutting edge will be in contact with the stone. In other words, when honing, hold it as you would a chisel. Rotate the bit as you move it back and forth over the stone. To hone the outside corners of the edge, lower the bit to as low as 20 degrees. (See Illus. 400.)

A spoon or a chair bit is held at a higher angle. Because the nose is upturned, the edge comes into contact with the stone when the bit is held between 60 and 70 degrees. (See Illus. 401.)

Polish the edge by moving the bit back and forth. At the same time, slowly rotate it. To reach the corners, lower the bit to as low as 20 degrees.

When you are done honing, examine the edge of your bit closely under a bright light. The polish created by the stone should extend all the way out to the cutting edge. If it does not, you did not hold the bit high enough while honing.

Once the bit is honed, take it to the buffing

*Illus. 400 (above left). When honing the outer edges of a spoon bit's cutting lip, hold the bit at about 20 degrees. Illus. 401 (above right). When honing the center of a spoon bit's cutting edge, hold the bit between 60 and 70 degrees.*

wheel. Hold the hollow side of the bit against the wheel so that the inside surface of the nose is polished. On my most commonly used bits, this surface gleams like a mirror. Now, turn the bit over and buff the outside. Duplicate the positions and movements used when honing. To buff the corners of the edge, hold the bit very low, raising the tang to bring the center of the edge into contact.

The bit should now be sharp. Examine it in a bright light by looking closely at (and directly onto) the edge. If you see any spots that glint, they are not sharp. Return the bit to the stone and the buffing wheel.

If you do not see any light reflecting when you examine the edge, test the bit on a piece of hard wood. Hold the bit in one hand and the wood in the other. Put the two together so that the bit is vertical to the surface of the wood and its nose is making contact. Push the nose into the wood, and at the same time turn it. If it is sharp, it will score a circle and pop loose a round chip. (See Illus. 405.)

I hone my spoon bits very seldom. Maintenance is done very quickly on the buffing wheel.

## USING A SPOON BIT

Before trying to use either a spoon or a shell bit in your woodworking, use it on a piece of plastic (such as PVC). The plastic has no grain, so the chip will not break apart. The core will hold together while you examine it closely, and you can even stretch it. This examination will help you to better understand how the bit cuts. (See Illus. 406.)

As the nose enters the wood (or plastic), it starts to cut a volute. It takes about one and a half turns to spiral this volute out to the bit's full diameter. (See Illus. 407.) After this, it slices a chip from the side of its round-bottomed hole. As the bit continues to turn, a plug (really a tight spiral) will start

*Illus. 402. Turn the bit over and buff the inside of the cutting edge.*

*Illus. 403. To buff the entire perimeter of the lip, roll the edge sideways from the center of the lip.*

*Illus. 404. I never grind spoon bits; instead, I hone them infrequently. Between honings, I restore their sharp edges on a buffing wheel. First, I buff the outside, using the same angles as when honing. Here, I am buffing the center of the lip.*

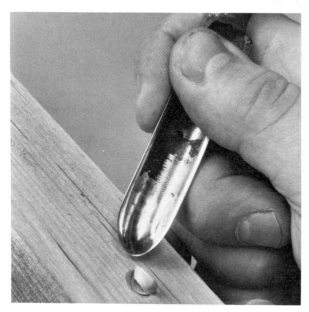

*Illus. 405 (above left). To test the sharpness of a spoon bit, push the lip into a piece of hard wood and turn it. The bit should score a circle and pop loose a round chip. Illus. 406 (above right). Try a sharp spoon bit in PVC. It will cut a spiral chip that looks like a piece of rotini pasta. Since the PVC has no grain, the chip will not break apart when you stretch it to examine it.*

Illus. 407. When you use a spoon bit, it first cuts a volute that spirals out to the bit's full diameter.

Illus. 408. The chip, in the form of a tightly packed plug, rises up the bit's hollow body.

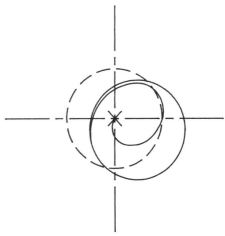

Illus. 409. If you start a spoon bit on a mark, it will spiral off center.

diameter away from the nose. (See Illus. 410.) Now, bore a hole. When you look at the top of the plug, you will see that the mark is centered. Practice boring more holes to learn exactly where to place the nose.

Because a spoon bit does not have a lead, it can

to rise out of the hole, guided upwards by the bit's hollow throat. (See Illus. 408.)

Before drilling the next hole, make a mark and try to start at this point. When you are done, look at the top of the chip; you will notice that the bit wandered and that the finished hole is just slightly off center. (See Illus. 409.) The mark is now near the edge of the spiral, rather than at its center. In thin work, such as a Windsor chair bow, this natural wandering can be a disaster.

To make a spoon (or chair bit) center on a point, shift the nose slightly. The mark should be on the hollow side of the body, about ¼ of the bit's

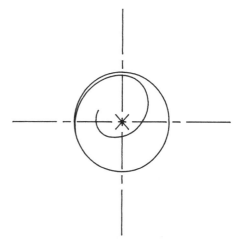

Illus. 410. Instead, start the bit ½ a radius away from the mark; this way, it will end up in the center of the volute.

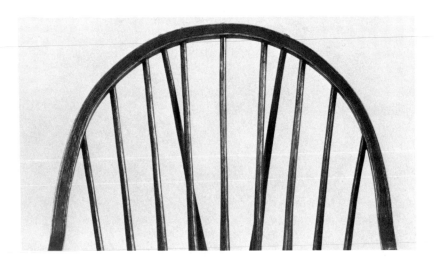

*Illus. 411. The spindles on this continuous arm Windsor chair intersect the bow at very acute angles. It would be difficult to make these holes for the spindles with most bits. However, spoon bits make these holes very easily.*

also walk across the wood's surface, damaging it. As a result, starting a spoon or a shell bit on a round part can be tricky. To prevent this, I set the nose where I want it and make a couple of quick, back-and-forth motions of the brace handle, moving it perhaps as little as ¹⁄₁₆th of a turn. This motion buries the nose into the wood, so that when I start the hole there is no risk of walking.

The lack of a lead makes it possible for a spoon or a shell bit to drill holes that are at very acute angles. The ability to change the direction of a hole after it is started is another important advantage when making a chair like a Windsor, that has sockets that are all at compound angles.

When you do drill a hole at an angle, remember this rule: spoon and shell bits have to be started square to the work. When drilling into a curved surface, such as a Windsor chair bow, start square to the tangent. Once the bit has begun to cut its full diameter (after the nose is buried), begin to pull the brace in the direction you wish while simultaneously turning the brace. The two motions cannot be done separately. Spoons and shell bits can easily cut very acute angles (such as those where the outermost spindles of a Windsor chair back meet the bow.) (See Illus. 411.)

These bits have no lead, and so are not pulled through the wood. The sharpened nose cuts because of pressure from your torso on the brace. This allows the bit to cut as rapidly (or as slowly) as you wish. Experience will teach you just how much pressure to apply.

Spoons and shell bits cut through holes equally well. Spoons do a better job of making blind holes. The upturned nose almost completely undercuts

the bottom of the compacted chip. The entire chip may not fall free of the hole, but you can easily dislodge it by tapping the part. The open nose of a shell does less undercutting. Breaking the bottom of the chip free from a blind hole may sometimes require that you reach into the hole with an object such as a nail.

A dull spoon bit creates a lot of friction and becomes quite hot. The cutting edge will not cut as aggressively and as cleanly. As a result, the chip will not hold together and breaks up into dust. Return the bit to the buffing wheel to restore the edge.

If your bit is not cutting aggressively, you may have to create more relief. Try grinding away some metal from the upturned nose, just under the cutting edge. (See Illus. 412.) This will allow the nose to bury itself further into the wood without riding on the curved forward surface.

One safety note: In a brace, a spoon bit is as dangerous as a mounted bayonet. Carry the brace so that you cannot fall or drive the bit into yourself. When you set the brace down, be sure that the bit is pointed away from you. This way, you cannot bump the cutting edge with your hand.

*Illus. 412. If a spoon bit is not cutting aggressively, it may need more relief. Carefully grind away some metal from under the upturned nose, as shown by the dotted line.*

# Appendices

# Appendix A:
# Tools for the Workshop

Following are lists of tools suggested for basic, intermediate, and advanced shops. These lists are intended as guidelines; they are not geared to your specific needs as a woodworker. For example, someone who makes wooden toys will need different tools from someone who makes furniture. Also, if you do very little moulding or rabbeting, you may choose to own a combination plane rather than special-purpose planes. Therefore, before buying any tools, think about the woodworking you will be doing.

Where I recommend iron planes in the lists, I have used Stanley's number system. A similar plane by another company will do just as well.

## Tools for the Basic Shop

1. *Bench Planes*
   A. Jack: wood, transition (#27), or iron (#5) with curved edge
   B. Smooth (for general smoothing): wood, transition (#24 or #35), or iron (#4)
2. *Moulding Planes*
   A. Bead: 3/16 inch
3. *Special-Purpose Planes*
   A. Block: 12 degrees (#9 1/2)
   B. Fillister: wood or iron (#78)
   C. Dado: 3/4-inch wood or iron (#39)
   D. Plow: wood, iron, or combination plane
4. *Combination Planes*
   A. #45 (will also replace some of the tools listed above)
5. *Plane-Related Tools*
   A. Wood spokeshave
   B. Drawknife (7 inches)
   C. Scraper: iron (#80)
6. *Saws*
   A. Hand: hand (7 points) and ripsaw (6 points)
   B. Backsaw: sash (12 points)
7. *Chisels*
   A. Firmers: 1/4, 1/2, 3/4, and 1 inch
   B. Gouges: out-cannel (1/2- and 3/4-inch regular sweep)
   C. Mortises: 1/4 and 3/8 inch

8. *Brace and Bits*
   A. Brace: 10-inch sweep
   B. Augers: #4, 6, 8, 10, 12, and 14

## Tools for the Intermediate Shop

1. *Bench Planes*
   A. Jointer: 24-inch (#7)
   B. Jack: wood, transition (#27), or iron (#5) with edge ground to a curve
   C. Smooth (for general smoothing): wood, transition (#24 or #35), or iron (#4)
2. *Moulding Planes*
   A. Bead: 3/16 and 1/2 inch, or combination plane
   B. OG: 3/8 inch
3. *Special-Purpose Planes*
   A. Block: 12 degrees, (#9 1/2)
   B. Fillister: wood or iron (#78)
   C. Iron rabbet (#92)
   D. Dado: 1/2- and 3/4-inch wood or iron (#39)
   E. Plow: wood, iron, or combination plane
   F. Match: 7/8-inch wood or iron (#48)
   G. Wood compass plane
   H. Forkstaff
   I. Nosing: 3/4 inch
4. *Combination Planes*
   A. #45 (will also replace some of the planes listed above)
5. *Plane-Related Tools*
   A. Wood spokeshave
   B. Drawknife (7 inch)
   C. Scraper: iron (#80 or #112)
   D. Toothing plane (if you do veneering)
6. *Saws*
   A. Hand: panel (10 points), hand (7 and 9 points), and rip (6 points)
   B. Backsaw: dovetail (15 points) and tenon (10 points)
7. *Chisels*
   A. Firmer: 1/4, 1/2, 3/4, 1, and 1 1/2 inches
   B. Paring: 1/4, 1/2, and 3/4 inch

C. Gouge: out-cannel (¼-, ½-, and ¾-inch regular sweep)

D. Mortise: ¼, ⅜, and ½ inch

8. *Brace and Bits*

A. Brace: 10-inch sweep

B. Augers: #4–#16

C. Center bits: ⅜, ½, and ⅝ inch

# Tools for the Complete Shop

1. *Bench Planes*

A. Jointer: 24-inch (#7)

B. Jack: wood, transition (#27), or iron (#5) with edge ground to a curve

C. Smooth (for general smoothing): wood, transition (#24 or #35), or iron (#4)

D. Smooth (for fine work): #4

2. *Moulding Planes*

A. Beads: ³⁄₁₆, ⁵⁄₁₆, and ½ inch or combination plane

B. OGs: ⅜ or ½ inch

C. Complex: assortment of at least three or four

3. *Special-Purpose Planes*

A. Block: 12 degrees (#9 ½) and 20 degrees (#60)

B. Mitre (if you use a mitre jack or mitre shooting board)

C. Rabbet: ¾ and 1 ¼-inch wood or iron (#180)

D. Fillister: wood or iron (#78)

E. Bench rabbet (#10)

F. Iron rabbet (#92)

G. Bullnose (#90 or #75)

H. Dadoes: ¼-, ½-, and ¾-inch wood or iron (#39)

I. Panel raiser

J. Plows: wood, iron, or combination plane

K. Match: ½ and ⅞-inch wood or iron (#48, #49)

L. Circular (#113) or wood compass plane

M. Side rabbet: wood or iron (#79)

N. Forkstaff

O. Nose: ¾ inch

P. Table leaf planes

Q. Sash fillister

4. *Combination Planes*

A. #45 (will also replace some of the planes listed above)

5. *Plane-Related Tools*

A. Wood spokeshave

B. Drawknife (7 inches)

C. Scraper: iron (#80 or #112)

D. Universal beader (#66)

E. Toothing plane (if you do veneering or work with heavily-figured woods)

6. *Saws*

A. Hand: panel (10 points), hand (7 and 9 points), and ripsaws (5 and 7 points)

B. Backsaws: dovetail (15 points), sash (11 points), and tenon (10 points)

7. *Chisels*

A. Firmer: ¼, ½, ⅝, ¾, 1, 1½, and 2 inches

B. Paring: ¼, ½, ⅝, ¾, and 1 inch

C. Gouge: out-cannel (¼, ½, ¾, and 1-inch regular sweep), and in-cannel (¼, ½, and ¾-inch regular sweep)

D. Mortise: ⅛, ¼, ⅜, ½, and ⅝ inch

8. *Brace and Bits*

A. Brace: 10-inch sweep

B. Augers: #4–#16

C. Center bits: ⅜, ½, ⅝, ¾, and ⅞ inch

D. Gimlets: ⅛, ³⁄₁₆, and ¼ inch

E. Spoons (assorted sizes if you want to make chairs)

# Appendix B: Bibliography

Dunbar, Michael, *Antique Woodworking Tools*. New York: Hastings House, 1977.

Dunbar, Michael, *Federal Furniture*. Newtown, Connecticut: The Taunton Press, 1986.

Dunbar, Michael, *Make a Windsor Chair with Michael Dunbar*. Newtown, Connecticut: The Taunton Press, 1984.

Dunbar, Michael, *Windsor Chairmaking*. New York: Hastings House, 1976.

Garvin, James L. and Donna B., *Instruments of Change*. Canaan, New Hampshire: Phoenix Publishing Co., 1985.

Goodman, W. L., *British Planemakers from 1700*. Suffolk (England): Arnold & Walker, 1978.

Goodman, W. L., *The History of Woodworking Tools*. London: G. Bell and Sons, LTD, 1962.

Hasluck, Paul N. (editor), *The Handyman's Book* (Reprint). Lancaster, Massachusetts: North Village Publishing Co., 1987.

Hummel, Charles F., *With Hammer in Hand. The Dominy Craftsmen of East Hampton, New York*. Charlottesville: University Press of Virginia, 1976.

Mercer, Henry C., *Ancient Carpenters Tools*. New York: Horizon Press, 1975.

Pollack, Emil and Martyl, *A Guide to American Wooden Planes & Their Makers*. Morristown, New Jersey: The Astragal Press, 1987.

Roberts, Kenneth D., *Tools for the Trades & Crafts*. Fitzwilliam, New Hampshire: Ken Roberts Publishing Co., 1976.

Roberts, Kenneth D., *Wooden Planes in Nineteenth Century America*. Fitzwilliam, New Hampshire: Ken Roberts Publishing Co., 1978.

Sainsbury, John, *Planecraft: A Woodworker's Handbook*. New York: Sterling Publishing Co., Inc., 1984.

Salaman, R. A., *Dictionary of Tools Used in the Woodworking & Allied Trades*. New York: Charles Scribner's Sons, 1976.

Sellins, Alvin, *The Stanley Plane: A History and Descriptive Inventory*. Augusta, Kansas: The Early American Industries Association, 1975.

Smith, Joseph, *Explanation or Key to the Various Manufactories of Sheffield*. South Burlington, Vermont: The Early American Industries Association, 1975.

Wildung, Frank H., *Woodworking Tools at Shelburne Museum*. Shelburne, Vermont: The Shelburne Museum, 1957.

# WEIGHTS AND MEASURES

| UNIT | ABBREVIATION | EQUIVALENTS IN OTHER UNITS OF SAME SYSTEM | METRIC EQUIVALENT |
|------|--------------|-------------------------------------------|-------------------|
| **Weight** | | | |
| *Avoirdupois* | | | |
| ton | | | |
|   short ton | | 20 short hundredweight, 2000 pounds | 0.907 metric tons |
|   long ton | | 20 long hundredweight, 2240 pounds | 1.016 metric tons |
| hundredweight | cwt | | |
|   short hundredweight | | 100 pounds, 0.05 short tons | 45.359 kilograms |
|   long hundredweight | | 112 pounds, 0.05 long tons | 50.802 kilograms |
| pound | lb *or* lb av *also* # | 16 ounces, 7000 grains | 0.453 kilograms |
| ounce | oz *or* oz av | 16 drams, 437.5 grains | 28.349 grams |
| dram | dr *or* dr av | 27.343 grains, 0.0625 ounces | 1.771 grams |
| grain | gr | 0.036 drams, 0.002285 ounces | 0.0648 grams |
| *Troy* | | | |
| pound | lb t | 12 ounces, 240 pennyweight, 5760 grains | 0.373 kilograms |
| ounce | oz t | 20 pennyweight, 480 grains | 31.103 grams |
| pennyweight | dwt *also* pwt | 24 grains, 0.05 ounces | 1.555 grams |
| grain | gr | 0.042 pennyweight, 0.002083 ounces | 0.0648 grams |
| *Apothecaries'* | | | |
| pound | lb ap | 12 ounces, 5760 grains | 0.373 kilograms |
| ounce | oz ap | 8 drams, 480 grains | 31.103 grams |
| dram | dr ap | 3 scruples, 60 grains | 3.887 grams |
| scruple | s ap | 20 grains, 0.333 drams | 1.295 grams |
| grain | gr | 0.05 scruples, 0.002083 ounces, 0.0166 drams | 0.0648 grams |
| **Capacity** | | | |
| *U.S. Liquid Measure* | | | |
| gallon | gal | 4 quarts (2.31 cubic inches) | 3.785 litres |
| quart | qt | 2 pints (57.75 cubic inches) | 0.946 litres |
| pint | pt | 4 gills (28.875 cubic inches) | 0.473 litres |
| gill | gi | 4 fluidounces (7.218 cubic inches) | 118.291 millilitres |
| fluidounce | fl oz | 8 fluidrams (1.804 cubic inches) | 29.573 millilitres |
| fluidram | fl dr | 60 minims (0.225 cubic inches) | 3.696 millilitres |
| minim | min | 1/60 fluidram (0.003759 cubic inches) | 0.061610 millilitres |
| *U.S. Dry Measure* | | | |
| bushel | bu | 4 pecks (2150.42 cubic inches) | 35.238 litres |
| peck | pk | 8 quarts (537.605 cubic inches) | 8.809 litres |
| quart | qt | 2 pints (67.200 cubic inches) | 1.101 litres |
| pint | pt | ½ quart (33.600 cubic inches) | 0.550 litres |
| *British Imperial Liquid and Dry Measure* | | | |
| bushel | bu | 4 pecks (2219.36 cubic inches) | 0.036 cubic metres |
| peck | pk | 2 gallons (554.84 cubic inches) | 0.009 cubic metres |
| gallon | gal | 4 quarts (277.420 cubic inches) | 4.545 litres |
| quart | qt | 2 pints (69.355 cubic inches) | 1.136 litres |
| pint | pt | 4 gills (34.678 cubic inches) | 568.26 cubic centimetres |
| gill | gi | 5 fluidounces (8.669 cubic inches) | 142.066 cubic centimetres |
| fluidounce | fl oz | 8 fluidrams (1.7339 cubic inches) | 28.416 cubic centimetres |
| fluidram | fl dr | 60 minims (0.216734 cubic inches) | 3.5516 cubic centimetres |
| minim | min | 1/60 fluidram (0.003612 cubic inches) | 0.059194 cubic centimetres |
| **Length** | | | |
| mile | mi | 5280 feet, 320 rods, 1760 yards | 1.609 kilometres |
| rod | rd | 5.50 yards, 16.5 feet | 5.029 metres |
| yard | yd | 3 feet, 36 inches | 0.914 metres |
| foot | ft *or* ' | 12 inches, 0.333 yards | 30.480 centimetres |
| inch | in *or* " | 0.083 feet, 0.027 yards | 2.540 centimetres |
| **Area** | | | |
| square mile | sq mi *or* m² | 640 acres, 102,400 square rods | 2.590 square kilometres |
| acre | | 4840 square yards, 43,560 square feet | 0.405 hectares, 4047 square metres |
| square rod | sq rd *or* rd² | 30.25 square yards, 0.006 acres | 25.293 square metres |
| square yard | sq yd *or* yd² | 1296 square inches, 9 square feet | 0.836 square metres |
| square foot | sq ft *or* ft² | 144 square inches, 0.111 square yards | 0.093 square metres |
| square inch | sq in *or* in² | 0.007 square feet, 0.00077 square yards | 6.451 square centimetres |
| **Volume** | | | |
| cubic yard | cu yd *or* yd³ | 27 cubic feet, 46,656 cubic inches | 0.765 cubic metres |
| cubic foot | cu ft *or* ft³ | 1728 cubic inches, 0.0370 cubic yards | 0.028 cubic metres |
| cubic inch | cu in *or* in³ | 0.00058 cubic feet, 0.000021 cubic yards | 16.387 cubic centimetres |

# INDEX